PROJECT MANAGEMENT FRAMEWORK

Project Management Framework

DAVID G. CARMICHAEL

The University of New South Wales, Sydney, Australia

A.A. BALKEMA PUBLISHERS / LISSE / ABINGTON / EXTON (PA) / TOKYO

Library of Congress Cataloging-in-Publication Data

A Catalogue record for the book is available from the Library of Congress

Typesetting: Grafische Vormgeving Kanters, Sliedrecht, The Netherlands
Printed by: Gorter, The Netherlands

ISBN 90 5809 325 5

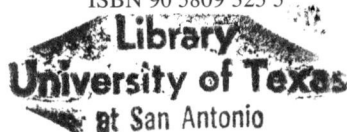

Contents

Preface

Project management, in many quarters, is 'flavour of the month'. So many people are jumping on the bandwagon and calling themselves project managers – the title has a nice sounding 'ring' to it. So many institutions and organisations are offering project management courses. So many people have 'sprung up overnight', claimed they are gurus of project management and spread the word across the land.

There is also developing a humorous side to project management such as in the descriptions:

> Project management is the art of creating the illusion that any outcome is the result of a series of predetermined, deliberate acts when, in fact, it was dumb luck. (Kerzner, 1989)

> Project management is the art of staggering as gracefully as possible between crises.

Concurrently with these events, there is an ongoing rational development of project management as a discipline.

This book covers the framework of project management. It offers a systematic development, and is intended to serve both teachers and practitioners. Project management material that is readily available in many texts has been deliberately omitted.

The book has developed out of lecture notes prepared for courses given to undergraduates, graduates and practitioners from a wide range of industries and backgrounds, as well as the writer's own experiences practising project management and interacting with project managers.

The book develops the material through looking at the fundamentals of projects and management, and then returns to a chronological development through the life cycle phases of a project. Examples are included throughout to reinforce the book material. The book applies a systems view to project management.

Few prerequisites are needed apart from a basic interest in projects, some exposure to the discipline, and an ability to think systematically and have an open mind. Given that projects are almost omnipresent, the first two prerequisites present few hurdles to most people, although the third may.

AIMS

The aims of the book are to understand project management and to contribute to project management thinking.

One of the overriding reasons for writing this book was to counter the myriad of misconceptions and thinking errors that exist among project management writers and practitioners. Project management, like management generally, introduces terms that people use any way they please, and pretend to be correct at the same time. Few people concern themselves with rigour in the usage of terminology; effort instead is put into impressing the receiver, much like the journalists' credo of 'Don't let the facts get in the way of a good story'. The way terminology is used in project management is not unlike:

> *'When I use a word,' Humpty Dumpty said in a rather scornful tone, 'it means just what I choose it to mean, neither more nor less'.*
> *'The question is', said Alice, 'whether you can make words mean so many different things'.*
>
> ('Through the Looking Glass', Ch. 6, Lewis Carroll)

Related views on the abuse of jargon, and faddish management thinking are contained in Carmichael (2002, pp.163-179).

As well, the level of thinking that goes into project management in many cases is very superficial and cookbook in nature. To counter this, the book adopts a systems view to provide a rigorous framework.

Management is shown to be a synthesis or inverse problem. As such, there are multiple solutions (decisions, or choices of control) possible. In most cases, managers are only after a satisfactory solution, or a solution that they can live with, and do not spend the additional time searching for the optimal solution. A manager may also be under time pressures to come up with quick solutions.

However managers expediently reverse the logic and deflect attention from their inability to come up with best solutions, on time pressures and pseudo 'practicality' arguments, when in fact managers do not understand the synthetic nature of their job. Managers are unaware of and do not understand the components of the synthesis problem, and so they never know where they are relative to the optimum. They are unable to vocalise or formulate the synthesis problem components; instead meaningless management jargon (Carmichael, 2002, pp. 163-179) is used as a smokescreen to hide their lack of competency. Such discussion goes to the very heart of current management knowledge being in its infancy, and current management education (read training) being superficial and low level.

ACKNOWLEDGMENT

The book contains numerous case examples contributed by as many people. Their contribution is gratefully acknowledged.

About the author

David G. Carmichael is a graduate of The University of Sydney (B.E., M.Eng.Sc.) and The University of Canterbury (Ph.D.) and is a Fellow of The Institution of Engineers, Australia, a Member of the American Society of Civil Engineers, formerly a Graded Arbitrator with The Institute of Arbitrators and Mediators, Australia and a trained mediator. He is currently a Consulting Engineer and Professor of Civil Engineering, and former Head of the Department of Engineering Construction and Management at The University of New South Wales.

He has acted as a consultant, teacher and researcher in a wide range of engineering and management fields, with current strong interests in all phases of project management, construction management and dispute resolution. Major consultancies have included the structural design and analysis of civil and building structures; the planning and programming of engineering projects; the administration and control of infrastructure projects and contracts; and various mining, construction and building related work.

He is the author and editor of seventeen books and over sixty five papers in structural and construction engineering, and construction and project management.

PART A

OVERVIEW

CHAPTER 1

Introduction

1.1 GENERAL

The original developments in project management are usually attributed to engineering disciplines particularly in the defence and construction industries. Today, project management is far wider than engineering projects. In a broad sense all people are involved with projects every day of their lives. This contributes to the plethora of people who call themselves project managers.

As well, many people either intuitively or deliberately convert management tasks associated with ongoing enterprises into a project management format – management by projects. For example in the management of a factory operation (mass production), a certain desired outcome and an associated time period are prescribed. This is then identified as a project and the full set of project management tools, including planning and control, are utilised to achieve the desired outcome. In the technical literature this has partly led to a blurring of the distinction between general business management and project management with reciprocal borrowing of ideas occurring. Extremely self-confident project managers accordingly claim that they can manage anything. The distinction between general business management and project management is an historical one and perhaps this will disappear with time. Project management's early push came from people with technical backgrounds. The push in business management is from people who might tend to use other parts of their brains compared to technically-oriented people.

Many people today term themselves project managers. Generally, such people are involved, at different levels and to different degrees, in the process of managing a project through its lifetime, and all will have different management styles depending on their personalities and circumstances. There are also those given the perceived prestigious title of project manager by their employers as a cynical trade-off for lower pay and lower work conditions, when other titles would be more appropriate.

In the community there is an impreciseness about the role and duties of project managers. This is compounded by there being no educational or vocational qualification requirement to calling oneself a project manager. In some contexts a project manager is not a new concept, only a new term.

There is similarly an impreciseness as to what constitutes a project. A number of definitions can be given but for now it is perhaps sufficient to think of a project as any undertaking with starting and ending points and with defined objectives as well as constraints. Research and development projects, for example, are different from construction projects, yet both share commonalities.

Projects may be large or small. Project start and end points may be different in different applications.

Project managers or members of project teams can face complex challenges in dealing with tight constraints on project resources, finances and the natural environment while attempting efficiency, productivity and performance. One of the characteristics of project management well practised is better control and use of resources.

It is often said that good project managers make it happen; poor project managers watch it happen.

Project management contains elements of both art and science. Project managers need competency in a number of areas. The analogy with decathlon athletes is sometimes made; there it is the aggregate score over a number of events that determines their success or otherwise.

1.2 PROJECT SUCCESS AND UNDER-PERFORMANCE

1.2.1 Influences and criteria

Through an examination of past projects, it is possible to identify some factors that are believed to contribute to project success, to project under-performance and/or project failure.

To some people, project success is profitability, or the profit margin left at the end of the project. On a resource project, success may be measured in terms of productivity, which translates into profit.

Project success, to others, is viewed in terms of 'beating plan', that is, beating the planned performance, typically in terms of time, cost and quality.

However performance measures in terms of time, cost and quality may not represent the total picture of project success. These measures do not necessarily include the views of all project stakeholders; the real test of success possibly is when all stakeholders believe the project to be a success. This might be in terms of satisfaction levels or opinions of the stakeholders, and as such is subjective and subject to influence. Often projects are 'sold' to stakeholders as being successful; the stakeholders' expectations are 'managed' (read manipulated). Measures of project success (and the frequency of measurement) are decided by the stakeholders or need to be agreed.

Each stakeholder may have, as well, its own internal measures of success. For example, project participants may have their own personal success measures such as improved status, enlarged experiences, training development, opportunities to have authority and responsibility, expectations of further work, perhaps more complicated, and improved ability/skills.

The useability of the end-product may also influence people's views on the success of a project.

Case example

Software development project

A software development project was initially deemed a success. The end-product was to the owner's specification and it was delivered in a timely and cost effective manner. The system fell into disuse within a year of its implementation. The project team knew what the problems would be and how to rectify them and had communicated this to the owner. The information loaded into the system and the catalogue definition used to categorise the information were both of poor standard. The project was considered a failure even though it delivered a product on time, within budget and to the required specification. The end-product was not useable and the project team was somewhat demoralised.

It is interesting to note that, by most people's standards, a project may be a failure but that it can later be rationalised into a success.

Case examples

Buildings

An internationally well known building cost approximately $3^1/_2$ times per square metre of floor to build more than equivalent buildings in the same neighbourhood.

Success was later viewed in terms of the building's looks, its lower running costs and more efficient operation, even though at $3^1/_2$ times the cost, the project would never have got off the ground.

As another example, a building for the arts was not completed on time or within budget. Success was interpreted in terms of its providing a world class venue for music and dance and the world-wide attention it gained.

Is it thus the case that a project can be a success despite poor project management performance? Or is it a case of confusion between what is the project and what is the end-product of a project? Or is it the case that the 'project' has been interpreted for project management purposes as the means of getting to the end-product, but for success/failure purposes the project is interpreted to also include the lifetime of the end-product? Clearly there is confused thinking somewhere here.

Case example

Consultant/contractor

One company views project success by both quantitative and qualitative means. The quantitative measure is profitability or overall profit margin remaining at the end of the project. This has to be considered in the light of the project risk involved. Project managers may not be in a position to influence initial profitability discussions – these, for example, might be dealt with by sales people.

The qualitative measures of success relate to the stakeholders' satisfaction with the project and include:
- Customer letters of commendation; customer confidence and faith; good customer relationships.
- Minimum defects, rework required, and issues that may resurface at a later date ('skeletons').
- New business opportunities made available or positive industry publicity.

These three measures may cause the success of a project to be retrospectively adjusted (upwards or downwards).

Skeletons cost the company time and money after the project is finished, and sour relationships.

Case example

Mining projects

Success in mining industry projects may be measured by the following factors:
- The project meets the (as-planned) performance anticipated by the owner in terms of it being completed on time, within budget and to the quality specification and requirements.
- The mine (incorporating the end-product infrastructure) shows a profit because of its ability to achieve planned productivity, at the planned cost of operation and at planned throughput rates, and to meet maintenance requirements.
- Satisfaction on the part of the owner and the operational personnel in that the mine infrastructure was completed, commissioned and handed over with the required training, operational documentation and maintenance specification, and with all modifications and rectification satisfactorily complete.

Under-performance

Project under-performance may be viewed like the negative of success. If the factors leading to project success are not followed, then under-performance could be expected to be the outcome.

Failure

Project under-performance may be referred to as failure by some people. Project under-performance implies a range (from no under-performance to heavily under-performing) while failure implies a sharp cut-off (either failure or no failure).

Learning

A consensus view of good practice is to become a learning organisation, whereby post mortem analyses of projects feed forward to practices on future projects. By this, project management is continuously improved, increasing the chance of project success and decreasing the chance of project under-performance. Better practice is to continuously note good and bad actions throughout a project rather than waiting till the end, when other issues may arise that prevent a comprehensive post mortem analysis, and memories have faded.

1.2.2 Factors affecting a project's outcome

The following is offered as a consensus view of the main categories of factors contributing to a successful project. It is emphasised that the factors and categories are only viewpoints, and are difficult to justify by any means other than they 'feel right' to many people. They are reasonable contentions but can't be proven.

In summary, the factors might be categorised as they relate to:
- The project manager.
- The project team.
- Stakeholders.
- Objectives and scope.
- Communication.
- Uncertainties/risks.
- Documentation.
- Early project phase work.
- Important matters.
- Alternatives.
- Planning.
- Control.
- Outsourcing.

These categorisations are not totally exhaustive.

A project contains an immense number of influencing factors (elements that contribute to bringing about any given result), none of which is possible to be isolated in order to establish its influence on a project's outcome. On any given project some factors may be present, some may not be present, while some factors may be more dominant than others.

It is not possible to demonstrate in any objective way how a particular factor contributes to project success. (In the terms of Chapter 16, the inverse investigation problem cannot be solved.) There is only experiential observation or anecdotes of projects to support popular contentions. To that could be added 'gut feel', judgement and perhaps logic; the contributions of various factors seem reasonable propositions. Experience says that the contribution of particular factors reduces/increases the chance of a project under-performing.

Objectivity is not possible because, although many projects are similar, no two projects are the same. A 'control' project cannot be set up, against which other projects with different factors are compared; the closest that may be obtained is to observe two similar projects but with different factors, but this is inconclusive. Each project cannot be done multiple times – each time with a changed factor. Medical experiments on people, it might be argued suffer the same problem, but get around the issue to a certain extent by conducting trials on large numbers of people. Conducting trials on large numbers of projects, each containing multiple factors that have the ability to influence the projects' outcomes, is not feasible. Individual factors cannot be varied one at a time, while all other factors are kept constant. There is not a one-to-one relationship between any project inputs/factors and project outputs; to a certain extent, a project (as a system) is almost uncontrollable in systems engineering (Kalman *controllability*) terms; altering any one input/factor affects multiple outputs. There are also *observability* (in the sense of Kalman) troubles. The benefits purported to be due to one factor may be due solely or partly to some other project factors.

Performing one project activity/task affects other activities, and the outcome of other activities. And on large projects there are large numbers of activities, most of which are interdependent.

An alternative is to approach the argument from the double negative perspective, namely by reasoning that with a certain factor, then project outcomes will almost certainly be undesirable. And this is supported by many case projects that have performed badly with such a factor. The consequent conclusion is drawn. Again this is reasonable, but can't be proven.

A number of studies have been reported trying to demonstrate a link between factors and project success. Typically they involve examining multiple real projects, and not any deliberate experimental projects. Projects are categorised, in a pseudo-rigorous fashion, as high, medium or low in terms of the factors and the (falsely drawn) conclusion then follows that there is a positive correlation between a factor and ultimate project success. The studies give an air of objectivity, but remain largely subjective.

The following discussion, within each of the success categories, is opinion.

The project manager

The selection of the right project manager who has personal drive and good leadership, people skills and communication ability is considered important. Some technical expertise is regarded as an advantage, as is a track record of delivering projects on time and on budget, and an understanding of the type of project. Assistance on some technical issues may be available from others.

The project manager is regarded as a key person. Poor project manager selection could be expected to lead to under-performance. The retention of good project managers within an organisation becomes an important concern.

A sufficient level of authority, along with agreed operational standards and guidelines, needs to be passed down from the project manager's parent organisation.

The appointment of the project manager early in the project is desirable.

Proactive behaviour is encouraged such that all eventualities are anticipated and there are a minimum of 'fires' that need to be put out. Reactive behaviour and 'crisis management' is to be avoided wherever possible. Trouble shooting and the ability to handle unexpected crises and deviations from plan are necessary.

High profile incidents, if managed well, can lead to an overall view of good management.

Commercial skills, particularly those related to managing a contract, are considered important.

On-the-job (OTJ) training together with attendance at formal courses is recommended practice.

Personal time, if not managed well, restricts the project manager from performing all tasks completely. This can be exacerbated by organisations under-resourcing.

Although working in a defined area, the project manager has to be capable of seeing the bigger picture, namely where his/her project fits into the overall scheme of the organisation and the likely calls on project resources.

The project team

One of the roles of the project manager is to develop team direction and a team wanting success or desiring accomplishment. This follows from careful selection of team members, with the appropriate expertise, clearly defined roles of team members, leadership, motivation, trust and mutual respect, recognition and reward. It is supported by achievable work goals, the provision of necessary information, resources, skills and authority, interest in the project, work challenge, group acceptance, and involvement in problem solving and decision making.

Motivation is seen as having a particularly strong influence on project success. It is required in all phases of a project. Where motivation is not present, it is seen to restrain project progress.

Within a team, good interpersonal relations are important.

Continuity of tenure of team members, either over the life of the project or parts of the project requiring applicable expertise, is seen as important. In any project team there is an investment of intangibles such as training, understanding, knowledge and coordination, which is lost when people are replaced. Loss of key team members can lead to the project under-performing while the learning process of the new people takes place. This can be reduced to some extent by appointing suitable replacements.

Team members may be lost for reasons internal or external to the project. Examples of the former include interpersonal conflict, and lack of suitable delegated authority. Examples of the latter include better rewards elsewhere, and redeployment within an organisation.

It is desirable that the project manager and project team be equipped with problem solving skills, have commitment, and not be subjected to a heavy workload, perhaps brought about by being involved in multiple projects simultaneously.

Stakeholders

A certain synergy is necessary among the stakeholders (all those influenced by the project), and in particular among the project manager, project team members, consultants, contractors, project owner and community. All have to work closely together. An agreed and well understood delineation of responsibility and authority is desirable, coupled with trust.

Misunderstood expectations or conflict between stakeholders could be expected to lead to under-performance. Misunderstandings have a tendency to blow out as the project progresses. Differences and disagreements between stakeholders can not only damage reputations and lose trust, but also lead to 'hard money' costs such as that which follows from replanning. Poor communications exacerbate the situation.

The project owner needs to clearly define its requirements and any relevant standards or performance criteria.

Stakeholders need to work in an atmosphere of openness and honesty.

Rapport is necessary between the project manager and the owner and the community, particularly if disputes are to be avoided.

Senior management in the project manager's parent organisation needs to demonstrate commitment by delegating sufficient authority and giving clear directions and priorities, thereby instilling confidence in the project team. A clear definition of responsibilities and clear communication channels between senior management and the project team assists here.

Failure to involve the end-user or disregarding the end-user's needs can result in conflict between the end-user and the project team.

Political influence and intrusion is to be avoided.

Objectives and scope

The end-product objectives and the project objectives are made clear to all stakeholders, some of whom may have contributed to their formulation.

There has to be a clear definition of the project scope, and this has to be understood by all stakeholders. All elements and processes of a project are interlinked. Treating each in isolation is dangerous.

Frequent scope changes and 'scope creep' can influence team morale, as well as lead to under-performance.

There is a need to properly define the project end point, final acceptance criteria and approving body, in order that the project just doesn't linger on.

Communication

The establishment of good communication links with project stakeholders and within the project team is seen as essential. This ordinarily includes the holding of regular meetings, two-way communication, and the sharing of knowledge.

Both formal and informal communication structures are necessary.

Uncertainties/risks

Attention is directed to the recognition of risks and the instigation of practices for dealing with these risks. These include, amongst others, risks arising from contractual, environmental (natural) and community sources, but typically anything likely to impinge on the time, cost and quality outcomes of a project.

An inability to assess risks, or lack of knowledge of risks, may lead to under-performance.

Unidentified risk events frequently come about. There is a consequent need to respond (belatedly and) adequately to the associated risks, otherwise project performance may be affected.

A lack of training of team members in risk management contributes to an inability to treat risks.

The impact of external forces such as changing technology, market shifts, politics, health, industrial disputes, and economy, need to be assessed early in a project. There are numerous examples of projects whose incompleteness stemmed from such causes.

Documentation

Documentation, including the design and contract documents, needs to be clear, explicit and well understood.

Directions and changes are recommended to be in writing.

Openness in stakeholders delineating their needs and expectations translates into more complete documentation, more realistic cost estimates and more realistic project timetables.

Early project phase work

Work carried out in the early project phases is believed to have the greatest ability to influence the project cost. Small expenses provide large returns.

It is suggested that the causes of under-performance occur primarily in the early project phases, and to a lesser degree later on where under-performance often occurs as a result of poor decisions made in the early project phases. There is a common lack of understanding of the importance of the early phases to the overall project outcome. Deficiencies in project management training of senior management contribute to this. Too often the early phases are rushed or given insufficient attention, leading inevitably to under-performance and commonly rework.

Specific activities that require special attention at the start of a project include feasibility studies, risk event identification, requirements definition, and responsibility establishment. Initial benefits plugged into feasibility study calculations may be overly optimistic and not achievable.

Important matters

According to the Pareto or 80-20 rule, 80% of the problems on a project arise from 20% of the items. Effort accordingly needs to be focused on the 20% of items (the so-called Pareto items). Focusing on the non important 80% of items may waste time and effort.

Activities on the critical path are examples of Pareto items.

Alternatives

There is always the potential for cost and time savings through the consideration of alternative work practices and using alternative suppliers. Such savings are commonly achievable because the initial project proposal may have been restricted by investment constraints when it was unsure whether the proposal would be accepted or not.

Generally this involves using systematic problem solving skills, though the more popular terms value management (analysis, engineering) or constructability (buildability) might be used to describe the same thing.

Planning

Recommended practice is to make use of established planning methodologies such as the critical path method, based on a work breakdown structure (breaking the project down into smaller units or work).

Poor planning could be expected to contribute to under-performance. Urgency, or the perception of the importance (or the felt need) to start the project as soon as possible, gets the project off on the wrong foot.

Planning is often not carried out because of the lack of training in planning skills of team members.

The ability to replan continuously throughout a project, adjusting to changing circumstances is important.

Coupled with planning is the task of developing realistic estimates. Estimates need to be conservative without being overly 'fat'. The conservatism acts like a contingency to take care of small unplanned expenses.

Control

The routine of planning-implementing-monitoring-reporting-controlling is carried out continuously throughout a project. Times, costs, resources and delays are monitored, variances (differences between actual and planned) analysed, and appropriate actions are taken. Common causes of variances are eliminated. Planned times, costs and quality are continuously reviewed and updated.

Lack of control or poor control could be expected to lead to under-performance as would not performing any planning-implementing-monitoring-reporting-controlling cycle components satisfactorily.

Planning has lesser value unless followed through with appropriate control actions.

A slow response to implementing controls can lead to a situation where it becomes too late to take effective control.

Mechanisms need to be in place to deal with changes in scope in order that any impact can be assessed and agreed. Continual changes of scope may confuse the project delivery process; there may even be an argument for stopping the first project and redefining a second project.

Outsourcing

Contractors and subcontractors (or internal organisational groups) of proven ability, and with the necessary financial and technical resources, are used.

Price should not be the sole criterion for selection. Working with preferred contractors is one way to go.

Case example

Selected causes of under-performance

The case example is described from the point of view of a company that designs, supplies, installs and maintains sophisticated electronic equipment in buildings. A number of reasons that can lead to under-performance in projects are cited. It may be difficult to determine one that is more important than the other. It depends on the type of project and difficulties faced within that project. Some possible solutions are given.

Lack of coordination/contractual arrangements

The sales team may agree to contracts with unrealistic requirements and expectations. Profit is trimmed, often making it difficult for the project manager to perform within the set restraints. Price variations to the contract may not be acceptable, causing difficulties when specifications are changed. Tight contracts may result in projects exceeding budget and deadlines. This situation arises not uncommonly. Sales incentives are in place for sales teams. Jobs are won and handed over to the design, supply, installation and operation teams. After handing over, sales generally do not have any further involvement. The project manager is required to manage the project, based on the criteria agreed to in the sales process, and is measured against these criteria.

Contractual obligations may place unrealistic expectations on the project manager. Parts of the contract may never be able to be achieved, for example, to provide details and prices of all products to be used in the project, and exact shipping details, that is, dates, name of vessel etc to the customer within 30 days of signing the contract. The customer may insist the contractual requirements be completed as initially agreed.

This situation arises in an organisational arrangement that does not combine the processes of sales, design, supply, installation and operation. The aim of the sales team is to win the job, and tender prices as accurately as possible. Sales people have clear financial targets in terms of the volume of orders they must attain in each quarter/year. They may provide a high level of customer service to the owner, but often the design, supply, installation and operation teams sort out the detail, and endeavour to bring the job in as tendered.

Part of the solution is to redefine the sales, design, supply, installation and operation processes, ensuring that others have input to the sales process, and the sales group have involvement after handing over to others. Thus, problems which are a result of the sales process are solved by a team involving people that were involved with that process. This encourages tighter ownership of the project.

Another part solution is, where a project involves a multitude of skills and products, it is desirable that senior management be involved with the development of the contractual conditions. For example, final approval for a project can be assessed by all groups involved, particularly when the project has high risks.

Inappropriate selection of project managers

A project manager needs many skills to successfully lead a project to completion. Not all project managers require technical skills, however, they do need commercial experience. In many cases, project managers are promoted from teams in which they have excelled technically. A project manager needs to plan, organise, lead, ..., and often requires a full time commitment. The importance of this is not always identified.

In some cases, project engineers are promoted to project management without adequate training in commercial areas and staff management. Senior management fails to identify the skills required in project management. A matrix based structure might not be used to draw in expertise in specific fields. These situations may arise from inadequate resources at commencement of a project, or promotion is given, without thought of overall requirements. Technically advanced project engineers do not necessarily make successful project managers.

A part solution is to run courses for project managers to improve the successful completion of projects. Project managers need support from senior management and a structure in place for assistance if required. Workloads may also be a problem and need reviewing – a project manager may not finalise all of his/her projects successfully, but if given half the number of projects, each could be successful.

Project analysis

Projects ideally should be analysed at completion to determine problem areas, and set up procedures to avoid recurrence. Some people continue to make the same mistakes in projects, and tend not to learn from their errors. For example, lack of planning by project managers can be easily solved by documenting practices, and making appropriate changes to improve the practices over time.

Incomplete specifications

Uncertainty regarding final requirements can lead to under-performance. Variations on deliveries, price, design work etc can cause delays. Specifications may not be complete until well into the design stage. Stringent requirements set out in contracts may make it difficult to accommodate changes.

When expectations are not clearly communicated at the front end of the project, careful negotiation with the parties may lead to positive outcomes.

Incomplete procedures

Incomplete procedures lead to costly mistakes and time delays.

For example, consider the packing marks required for an export project. There may be no documented procedures to follow, or a person to take responsibility. The packing process may be done in a short time frame, leaving it open for error. Incomplete procedures result from a lack of forward planning and can be solved by identifying these issues at the front end and finalising details before the due date.

Resources

Team members may be chosen due to availability rather than expertise or experience. A project team must work closely to achieve required performance. Without appropriate training and leadership, work may not be carried out to required expectations, delaying the project. In tight project situations, time constraints do not allow for retraining and development, and the team member is left to cope unaided. Within the company, employees may be moved from job to job. When cross divisional transfers occur, more problems may be experienced due to the different cultures of the businesses. These problems are difficult to solve. The company can use more subcontract consultant labour, but immediate familiarity with the company's practices may be lacking. Alternatively the company might invest more time in job orientation.

Cultural differences

Cultural differences can lead to a misinterpretation of requirements. In some countries, it is necessary to build up trust, respect and friendship with the stakeholders to enhance positive progress in a project. Cooperation improves when this ground work is complete. Building relationships can be used positively in influencing the decisions of the stakeholders. Incorrect assumptions on technical details or desired requirements leads to confusion and frustration, and more importantly, time delays. Positive approaches and interactions assist with cooperation.

In a climate of cultural differences, information may be classified as:
• Issues discussed openly by the parties.
• Withheld information.
• Information not usually discussed.
• Information that may emerge but was unknown by the parties.

Success or under-performance may be directly linked to the amount of information available to the project manager.

1.3 BOOK OUTLINE

The book develops the subject through five parts:

Part A – Overview
This part covers the fundamentals of projects and project management. And taking the view that management involves solving problems and that management problems are synthesis/inverse problems, suitable discussion is given on systematic problem solving and the systems engineering approach.

The next two parts of the book are presented in terms of a chronological development through the life cycle phases of a project.

Part B – Starting a Project Off
This part covers things that are peculiar to starting off a project, and generally don't occur again in the project.

Part C – Other Life Cycle Activities
This part gives a summary of the activities remaining in the life cycle of a project. Such activities constitute the bulk of the many available project management texts, and hence are only mentioned in passing in this book. Activities are listed according to:
• The remainder of those things that are peculiar to starting off a project, and generally don't occur again in the project.
• The recurring issues within the body of projects. They give rise to synthesis problems with ever increasing (evolving) detail available.
• Those things that are peculiar to finishing off a project and generally don't occur earlier in projects.

Part D – Project Examples
A number of interesting projects and issues considered from a project management perspective are given in this part. Some less well known project examples are given.

Part E – A Fresh Look at Project Management
This part gives a new way of looking at project management, a view of project management not available elsewhere.
    Examples are included throughout to reinforce the subject material.
Figure 1.1 shows a schematic outline of the flow of the book.

Terminology uses

Terms used interchangeably by writers include owner, client, principal, employer, developer, proprietor, purchaser. Generally, the first term is the one adopted in this book.

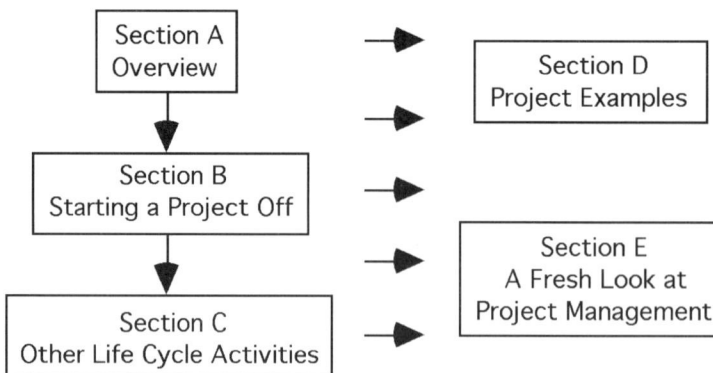

Figure 1.1 Outline of book.

Gender comment

Commonly references use the masculine 'he', 'him'', 'his' or 'man' when referring to project personnel possibly because the majority of project personnel have historically been male. However such references should be read as non-gender specific. Project management is not an exclusive male domain.

EXERCISES

1 When was the term 'project management' first used?
   What was the terminology evolution leading to the usage of this term?

2. How do you measure the success of a project? What do you have to do to ensure a successful project? What are the factors that most influence a successful project outcome?
   What are the causes of under-performance in projects? Rank these causes according to your opinion and justify your ranking. How do such causes arise? How might these causes be eliminated?

3. Listed above are a number of factors contributing to project success or under-performance. It may be difficult to determine one that is more important than another. It depends on the type of project and difficulties faced within that project.
   Select the ten most important factors as you perceive them affecting the success of your projects. Rank these factors. Justify your ranking.

4. On a project in which you are involved:
   • Identify the causes of under-performance.
   • Establish how these causes arise.
   • Develop means of eliminating these causes.

5. On a project in which you are involved:
   • Identify the factors by which the success of the project is judged.
   • Identify what is necessary in order to ensure a successful project.
   • Determine the most important of these.

6. In some projects, budgets may not be identified up front, little data may be available for planning, historical information may be lacking etc. In such cases, how does a project manager ensure a successful project completion?

7. Some people refer to key performance indicators (KPIs) as a means of establishing whether a project is successful or not. What might be typical KPIs on a project? What is the connection between project objectives and key performance indicators?

8. Some people divide project descriptors into hard (tangible, quantitative, ...) and soft (intangible, qualitative, ...) groups. Examples of the former include anything that can

be measured, such as time and cost. Examples of the latter include cooperative and positive attitudes, commitment to the project, and ethical conduct.
How would you relate such an approach to a discussion on project success and under-performance?

9. What influence do previous similar projects play in gauging whether a current project is successful?
In some industries, work is done a certain way today because this is the way it was done yesterday. How does such an attitude or culture influence your view of project success?

CHAPTER 2

Projects and Project Life Cycles

2.1 INTRODUCTION

Project management is a very broad subject and accordingly there are many ways that you can start a discussion on the subject. Perhaps the most fundamental way to start a discussion is to first examine what it is that is being managed, namely what is a project.

2.2 WHAT IS A PROJECT?

2.2.1 Definitions

A reasonable place to start any discussion on project management is to first examine what is meant by a project. Dictionary definitions do not shed much light on what is a project. The definition,

Any undertaking, or set of activities (tasks), with starting and ending points, and with defined objectives and constraints, and resource (people, materials, equipment) consumption.

is more useful, though still perhaps unsatisfactory in a number of ways. Most published definitions are not much use and many (even those in some project management texts) are not correct. Most people have their own idea of what constitutes a project without necessarily being able to put it into words. It is not unlike trying to define an elephant – everyone knows an elephant when they see one but has trouble developing a comprehensive definition. So, finding a satisfying definition of a project is difficult.

2.2.2 Attributes

Perhaps a more suitable way of thinking about a project would be in terms of attributes that are characteristic of projects, including that they are:
- Unique (one-off, specific discrete undertaking with a unique environment and unique constraints).
- Finite (definable start and end points).

and also, though not exclusively:
* Multi-disciplinary.
* Multi-organisational.
* Complex.
* Dynamic.

The uniqueness attribute is not questioned for significant projects like a space mission or the construction of a major edifice. However, there are projects which are only slightly different to projects that have gone before. Examples include the construction of an 'off-the-shelf' house, and getting yourself to work each morning. The construction of an 'off-the-shelf' house might have a different owner, site conditions and weather leading to different constraints, but whether this gives it sufficient uniqueness is arguable.

The finiteness attribute is unarguable. The multi-disciplinary, multi-organisational and complexity attributes are not applicable for small projects, such as getting yourself to work each morning, but are applicable for large projects. The degree of dynamism of projects is relative to ongoing enterprises which might be thought of as repetitive and quasi-static from one day to the next.

Examples

Clearly projects can be of any size ranging from something as small as 'going shopping' to something of much grander proportions such as a 'space mission' or the construction of a large bridge or dam.

Clearly, also, projects have always existed. Most people are aware of major projects such as the construction of the Great Wall of China, the pyramids of Egypt and the Suez and Panama canals. The undertaking of such projects would have posed many management problems.

Interestingly though, the term 'project management' is only several decades old and possibly arose from the greater present day demands of economic pressures, resources, competition and the welfare of the project workforce. New attitudes and methods, developed in response to these demands, constitute the emerging discipline of project management.

Projects exist in all spheres of human activity. For example:

Civil engineering, construction, petrochemical and mining projects involve site-oriented activities often remote from a head office. Exposure to the weather and the possible remoteness of the site introduce unique risk sources and require special organisation and communication skills. Such projects also tend to be capital intensive.

Manufacturing projects may be for special purpose items or for mass produced items. The owner may be one-off or the community at large. The projects tend to be carried out in a factory where the environment can be controlled, though installation and commissioning may be remote from the head office.

Management projects involve the organising and coordination of people with little or no identifiable physical end-product.

Research projects may have desired end-products that might not be reached; they may be long term and operate in a grey area of knowledge. Milestone events may be hard to identify and thus render management difficult.

Generally, however, each project is different and the steps taken to go from idea to completion are different between projects. As well, projects have a dynamic nature about them.

2.2.3 Projects and subprojects

A given project may have many project managers. It is suggested that whilst this is confusing when looking at the big picture of a major project, such as the laying of a cross country pipeline, when looking at the activities of each participant within the project there can again be a set of projects (equivalently subprojects, or subsystems within a system).

To expand on the problem using the cross country pipeline example: the pipeline owner is concerned with everything from the first glimmer of an idea that leads to a feasibility study to the completion of construction and commencement of operation (and maybe even including the operation for its useful life). This view includes every conceivable activity necessary to see the project completed. The pipeline owner may have a project manager within its own company to look after the project. The pipeline owner may then employ a project management company whose brief is constrained; it may not include the feasibility study, the project financing, the undertaking of the physical work, the responsibility for survey and many other activities. However, the project management company appoints a project manager. This second project manager now organises the work into

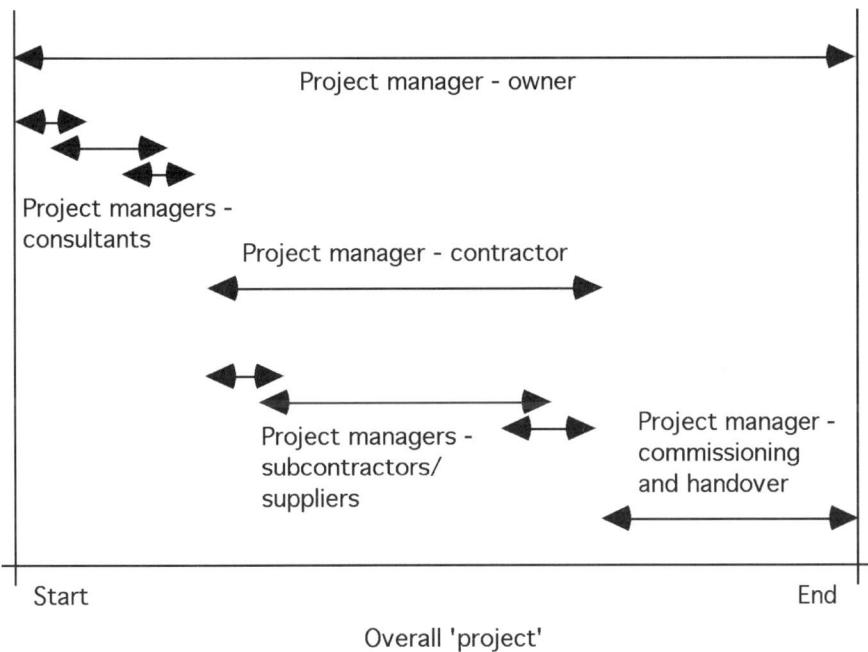

Figure 2.1 Timespans of influence of various project managers on a major project.

smaller sections to be undertaken by other companies, say a construction contractor or insurance broker. These other companies can also appoint project managers and so on down the line.

The important point here is that each organisation is correct in its use of the term project manager. In each case there is a scope (of work) that has a defined start, finish, need for resources and results in a finite end-product. The fact that each end-product is only a small component of a larger end-product or contributes towards a larger end-product or project is immaterial to the organisation making the component. Thus the view of a project and the responsibilities of the project manager are relative to the responsibilities of the organisation undertaking the project (Fig. 2.1).

The foregoing situation of course results in a multiplicity of project managers on a given major project. Such multiplicity can cause considerable confusion. Rarely do project owners attempt to resolve this confusion such as by requiring, as a condition of contract, that specified titles be used throughout the project in the course of all correspondence and discussion. This is possibly because superficially it may be successful, however when you look more closely you may find that the title project manager is still being used informally within the various organisations making up the participants of the project.

In the end result, whether a person is a project manager or not is purely relative to the view that that person or that person's organisation takes of the project.

This multiplicity of project managers may perhaps be better viewed from a systems viewpoint. A project is a system and a system can be decomposed to interacting subsystems which are themselves systems. Likewise a project can be decomposed to interacting subprojects (Fig. 2.2). Each subproject, which is itself a project, looks after a part of the total project and is overseen by a person with the title project manager. More correctly these people should be referred to as subproject managers but such a title does not sound as important.

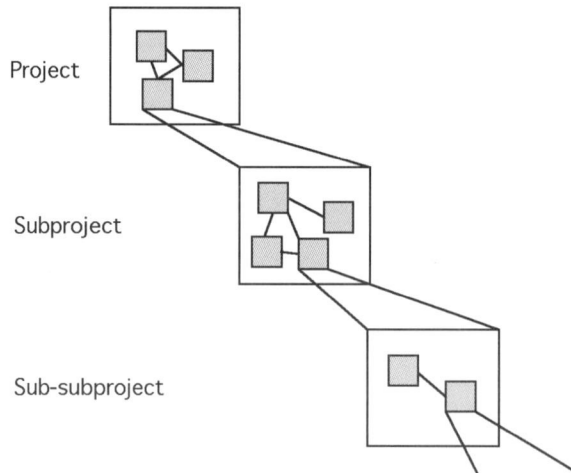

Figure 2.2 Decomposition of a project.

Example

Consider a building project:

- From the owner's viewpoint, the project is everything that brings the building into being. The owner engages a project manager from within the owner's organisation (in-house) or as an external consultant (outsource). This project manager has a wide span of responsibility for the project work and project control, and has involvement from the concept of the project through to its completion. The appointment and involvement begins before other project participants begin and lasts till after they finish. Such a person represents the interests of the owner and may have a certifying role.
- The owner (or owner's project manager on the owner's behalf) engages consultants such as an architect and structural engineer to carry out work such as the building design. To each of these consultants, the project of concern is the particular design aspects of the building. Each consultant has a project manager to oversee the particular design.
- The owner (or owner's project manager acting on the owner's behalf) engages a builder to do the on-site construction work. To the builder, the project of concern is the construction of the building. The builder has a project manager who looks after costs, quality, time, and contractual and related matters. Such a concept of (project) management is very old.
- Other participants that contribute to the larger project include other consultants and tradespeople. Each works within its own subproject.

Occasionally one of the consultants may perform a dual role, for example, as project manager for the owner, and as a design consultant, although this can introduce a conflict of interest (Carmichael, 2000).

2.2.4 Programs

The term program is used to describe aggregates of projects, just as projects are aggregates of subprojects. Alternatively, subprojects are subsystems of projects, and projects are subsystems of programs.

Within an organisation, there may be prioritisation issues between projects, project team rivalry and competition for resources.

2.3 PROJECT PHASES

2.3.1 Changes

The application of resources (people, materials, equipment,...; money) varies throughout a project, perhaps in a manner shown qualitatively in Figure 2.3. Initially project progress may be slow as the project team, resources and the work become organised. Momentum is maintained during the middle phases of a project. Near the end, progress tends to slow

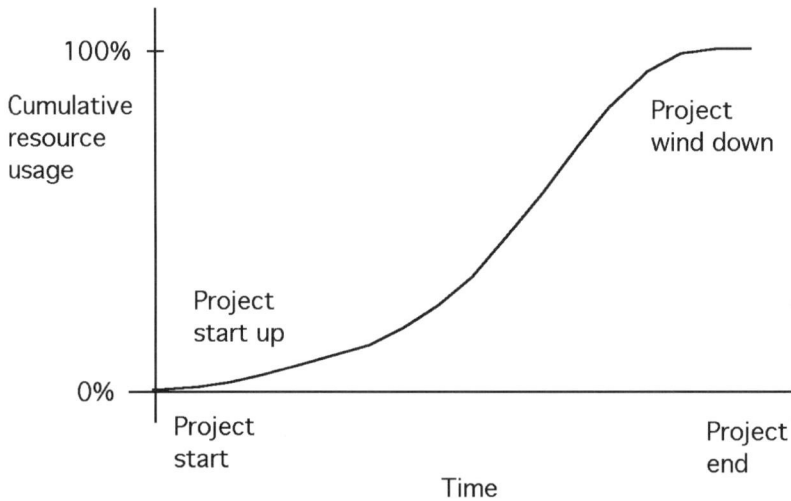

Figure 2.3 Example resource usage over a project's life.

down and this is the case with most projects. For example, observe the progress of a new neighbourhood house being built. The different work and resource levels required in the different phases place special requirements on the associated management. The duties of the project manager and the project team change throughout the project lifetime.

This is to be contrasted with ongoing businesses where, once established, there is a relatively constant level of resources and resource type required. The duties of a manager accordingly change little compared with project management duties.

Special management techniques have been developed to cater for the idiosyncrasies of projects.

2.3.2 Phases

Based on observation, projects are seen to be separable into phases. Generally the phases will occupy different time spans. Different writers give the phases different names but the intention is the same. The naming changes with the type of project. There is no definitive naming of phases. There is no definitive number of phases. Some management texts consider four generic phases:
- Initiation (concept, ...).
- Development (planning, ...).
- Implementation (execution, construction, ...).
- Termination (finishing, completion, ...).

See for example (PMI, 1987):

Concept (Conceive)
- Gather data
- Identify need
- Establish:
    - goals, objectives
    - basic economics, feasibility
    - stakeholder
    - risk level
    - strategy
    - potential team
- Guesstimate resources
- Identify alternatives
- Present proposal
- Obtain approval for next phase

Development (Develop)
- Appoint key team members
- Conduct studies
- Develop scope baseline:
    - end-product(s)
    - quality standards
    - resources
    - activities
- Establish:
    - master plan
    - budget, cash flow
    - work breakdown structure (WBS)
    - policies and procedures
- Assess risks
- Confirm justification
- Present project brief
- Obtain approval to proceed

Implementation (Execute)
- Set up:
    - organisation
    - communications
- Motivate team
- Detail technical requirements
- Establish:
    - work packages
    - information control systems
- Procure goods and services
- Execute work packages
- Direct/motivate/forecast/control:

         - scope
         - quality
         - time
         - cost
      - Resolve problems

Termination (Finish)
       - Finalise product(s)
       - Review and accept
       - Settle final accounts
       - Transfer product responsibility
       - Evaluate project
       - Document results
       - Release/redirect resources
       - Reassign project team

Some writers choose to omit an initiation phase, perhaps including this as a *strategic phase* determined by the corporation or business responsible for instigating projects. Some writers also choose to include an additional phase, an *asset management phase* at the end, though most regard the project as complete before the maintenance and operation management work takes over (although the definition of a project is sufficiently flexible to allow this); also complete asset management work begins in the conceptual phase of project management.

The choice of what constitutes a phase may depend on such matters as a time division, budget division, resource type usage or management structure. The argument can be reversed such that a pre-established phasing determines, for example, the resource type employed or the time allocation.

The choice of the number of phases is similarly open to different interpretations. There is no consensus on how many phases a project should have or what they should be called.

The project phases may be subdivided into *stages*. Note though that a number of writers use the terms phase and stage interchangeably.

Example – project management services

Contracts for project management services may list services provided by a project manager under project phases. The parties to the agreement agree as to what services and the appropriate extent of services the project manager should provide. For example (WAPMA, 1990):

Project Inception Phase
• Site use studies
• Zoning analysis and authorities consultations
• Finance and feasibility studies and reports
• Project programming and analysis

- Building procurement advice
- Consultants
- Advising on site selection
- Studies of services
- Soil investigations
- Existing building survey and reports
- Submissions for grants, subsidy, fund raising
- Phased development planning
- Special studies
- Submissions; attendances
- Litigation and arbitration

Project Design and Construction Phase
- Schematic design
- Design development
- Contract documentation; contract administration
- Cost planning
- Time planning
- Segregated contracts
- Landscape design
- Furniture, furnishings and art works
- Construction management
- Examination of drawings and documents

Project Handover and Commissioning Phase
- Handover
- Commissioning assistance
- Post construction documentation
- Defects liability period
- Final accounting
- Final certificate
- Services after final completion
- Estimates of replacement costs
- Reinstatement after damage
- Property estimates

Example – Equipment acquisition

The equipment acquisition process may follow the following phasing:
- *Need identification*; budget approval.
- *Exploration of alternatives*; equipment data compilation and equipment demonstration.
- *Selection*, or development/production of new equipment.
- *Deployment* of equipment.
- *Operation, maintenance and support* (or post-project).

Example – Movie making

Movie making effort can be divided into phases where each phase represents the completion of activities/tasks, or relates to decision points, particularly whether or not to proceed to the next phase.

Possible movie project phasing terminology is:
- *Initiation.*
- *Pre-production.*
- *Production.*
- *Post-production.*

The phases loosely reflect a generic chronological model for project management. In the initiation phase, the decision to go forward with the film would be made – there is no point in proceeding to conduct pre-production activities if there is not a good film idea or funding is not secured.

There may be overlap between phases, and different films, depending on their origin and nature may interchange phases in which certain activities are carried out.

Example – Pharmaceutical product development

Phasing for new pharmaceutical product development, conducted within any statutory constraints, may take the form of:
- *Pre-clinical testing*; selection of sample and test regime.
- *Clinical testing*; selection of sample and test regime.
- *Registering product*, patenting product and name; government approval.
- *Production and marketing.*

The timescale for this product development may be over a number of years.

Example – Organisational change

Change may be considered to go through definable project phases.

Initiation
There would be a perceived need for change. The project objectives and scope follow. Responsibility to develop and implement the changes is delegated to a project manager. Alternatives, costs, legal implications etc of the change are considered. Advice and expertise is sought to develop plans and policies to implement the change.

Development
A project team, possibly comprising external consultants and key people who will be involved in implementing and maintaining the change, is organised. Lines of authority are established. The scope is better defined. Responsibility for people with professional or technical expertise is defined.

The change process is planned, outlining the key activities and expected durations, resources and the approvals and hold points. The plan of action is approved by senior management.

Implementation
Senior management support in terms of finance, and the development of new policies and procedures progress the change. The status of the project is monitored and reported with reference to the approved plan of action. Scope changes, quality and costs are also reported.

Termination
The termination of the project can be defined as whenever convenient. For example, it may be defined as when the new operations or procedures are established, even though there may be an ongoing responsibility for maintenance. This occurs with new safety regulations, for instance, where ongoing site safety inspections and new safety plans, policies and procedures have to be maintained.

Example – software development

A phase description for a computer software project might be:
* *Project definition.*
  (Feasibility compared to competing software.)
* *Business specification.*
  (Required functions, interfaces, performance.)
* *Design.*
  (Specification, architecture, data structure, interfaces, algorithms.)
* *Code and test.*
  (Program, test plan.)
* *User acceptance test.*
  (Test plan.)
* *Documentation.*
  (Manuals.)
* *Implementation.*
  (Installation, integration, functioning, training.)
The software development industry has its own methodologies and terminology.

Example – chemical process plant construction

The following gives a breakdown of activities for a chemical process plant project. The activities have been divided into Pre-Investment and Investment phases (UNIDO, 1978).

A. *Pre-Investment Phase*
   1. Project Scope
      1.1   Product and raw material specification.
      1.2   Plant capacity including standby and future capacity additions.
      1.3   Type of plant – modular or block construction.
      1.4   Operating and control requirements
      1.5   Location and site.

2. Project Development
2.1   Process selection and evaluation: Distillation v other separation processes.
2.2   Process acquisition. Licensing agreements, if any.
2.3   Project Feasibility Studies including project financing and environment (natural) considerations.
2.4   Board approval to proceed.
2.5   Tender stage.
    (i)   Tender document preparation including technical specification and commercial conditions.
    (ii)   Selection of approved contractors.
    (iii) Tender evaluation and negotiation.
    (iv) Contract award.

B. Investment Phase
3. Project Implementation; Owner and Contractor
3.1   Kick-off meeting between owner and contractor.
3.1   Review of pre-contract documents and contract.
3.2   Process and engineering data supplied by the owner.
3.3   Project program.
3.4   Cost control and accounting procedures.
3.5   Approval procedures.
3.6   Progress meetings and reporting.
3.7   Correspondence.
3.8   Change and variation notice procedures.
3.9   Project numbering system.
3.10 Preferred suppliers.

4. Project Initiation
4.1   Project manager and project team.
4.2   Review pre-contract documents.
4.3   Scope of work and services.
4.4   Process design basis.
4.5   Administrative procedures, accounting, progress claims and approval procedures.
4.6   Reports and documentation.
4.7   Manuals and training.
4.8   Review engineering plan.
4.9   Engineering work scope.
    • Design specifications and standards.
    • Use of computer-aided-engineering capability.
    • Completion date.
    • Future capacity expansion.
    • Utility facilities.
    • Product and raw material transport.
    • Spare parts and equipment.
    • Operating and maintenance procedures.

- Standby capacity.
- Requirements for process and service buildings and enclosures.
- Equipment spacing and layout.
- Communication requirements.
- Philosophy for isolation and contamination. Industrial hygiene.
- Local ordinances and approvals.

4.10 Review procurement plan.
- Critical items of equipment and materials.
- Free-issue items and materials.
- Preferred or sole suppliers.
- Quality control and inspection procedures.
- Number of competitive bids.
- Purchase policy, overseas or local.

4.11 Review construction plan.
- Site data and local area survey.
- Site preparation, site access and weather restrictions.
- Construction methods, requirements and erection sequence, tall lifts.
- Site fabrication vs shop fabrication.
- Temporary construction facilities.
- Heavy rigging.
- Site receiving requirements and facilities.
- Construction staffing and organisation.
- Field inspection and quality control.
- Field safety procedures.
- Labour availability and hiring procedures.
- Labour contraction camps.
- Local codes and permits.
- Construction planning, scheduling and cost control.
- Union attitudes.
- Hours of operation.
- Safety and hazards.

5. *Process Engineering*
- Establishment of design basis.
- Basic engineering including:
Material and energy balances.
Process flow diagrams.
Equipment data and specifications.
Piping and Instrumentation diagrams.
Utility requirements and distribution.
Plant layout.
Electrical distribution.
Effluent and emission specifications.
Operation manuals.
Maintenance manuals.

6. *Project Engineering*
   • Detailed engineering of the plant.
   • Identification of time and process critical equipment items.
   • Pre-qualification of vendors for the supply of equipment and materials.
   • Procurement of all equipment and materials for the plant.
   • Inspection of equipment and materials during fabrication, on delivery to site.
   • Provision of test certificates for equipment and materials where required.
   • Packing and transport of equipment.
   • Arranging insurance cover.
   • Purchase, clearing, levelling and otherwise developing the site.
   • Testing soil characteristics of the site, particularly areas for heavy loads.
   • Construction of roads within battery limits.
   • Arranging communications on site and to head office.
   • Design and construction of all civil work on site.
   • Receipt and inspection of equipment on site and storage.
   • Provision of all erection equipment, tools and cranes.
   • Erection of all equipment and piping.
   • Providing training for all plant engineers and operators.
   • Testing of all erected equipment individually, by sections and as a plant and carrying out all pre-commissioning procedures.
   • Supplying all feedstock materials and all other inputs necessary for the start-up of the plant.
   • Commissioning and start-up of the plant.
   • Operation of the plant from start-up until completion of the performance guarantee tests for the plant.
   • Demonstrating the performance guarantee tests.

Example – Bank projects

One bank's project management manual outlines the broad phases that its projects go through:
• *Start up* – encompassing project proposal (intent, scope, milestones, stakeholders, risk, endorsements), project justification, project planning, establishing measures against which the project will be monitored, and project approval.
• *Planning and specification* – consisting of design, specification, and implementation planning.
• *Build and implement* – this includes construction, implementation, and training.
• *Completion* – this handles project closure and benefits tracking.

All of these phases are tied together by continual project management monitoring and reporting.

Example – Building projects.

An example phasing of building projects is (Institute of Building, 1979):

Initial, with the [owner]
- Project managers' brief
- [Owner] involvement
- Viability of project
- Funding
- Communication and reporting
- Grants

Feasibility
- Develop brief
- Project feasibility
- Program
- Design team
- Site
- Government approval
- Site investigation
- Preliminary drawings
- Budget
- Outline planning approval
- Feasibility report

Pre-construction
- Contractor selection
- Management structure
- Communication
- Program
- Design proposals
- Monitor budget
- Costing
- Pre-ordering
- Accounting
- Planning permission, Building Regulations
- Contract documents
- Contractor prequalification
- Selection
- Tender evaluation
- Site inspection

Construction
- Monitor progress
- Hoarding and existing site conditions
- Meetings
- Inspect works
- Certify payment
- Safety

- Quality control and testing
- Anticipate
- Monitor budget and variation orders
- Training
- Certify
- Statutory undertakers

Completion
- Pre-commission
- Final account
- Commission
- Manuals
- Defects liability period
- Feedback

Example – Market research projects

Typically, market research involves:
- *Situation analysis.*
- *Formal investigation.*
- *Analysis and interpretation of data.*
- *Reporting.*

2.3.3 Phase transitions

Divisions between phases are not necessarily characterised by noticeable discontinuities in project inputs or outputs. Divisions may be conveniently chosen to match, for example, the approvals or documentation requirements of a project. Owners like to be informed at significant milestone points in a project and to review the progress of the project; whether the project continues unchanged, continues in a modified form or is abandoned will depend on information supplied to owners at these milestone points. Such points are convenient division points for a project and hence divisions between phases.

Following the end of one phase, the plan for the succeeding phase unfolds and can be developed in detail. A broad plan, of course, would have been developed at the start of the project.

Trouble with using the idea of phases comes about when it is recognised that there can be a blurring of activities across phase boundaries. All project activities are not sequential or serial in nature; feedback occurs between activities as ideas and information are consolidated.

The idea of phases nevertheless, provides a useful way of distinguishing between an ongoing enterprise and a project. With an ongoing enterprise it is possible for all phases to be active at any one time, while a project progresses from one phase to the next.

A project is a system. A system can be broken into subsystems with interaction between the subsystems. Each subsystem is itself a system which can be further broken down. A

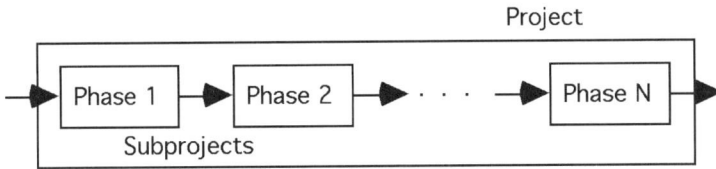

Figure 2.4 Project phases as subprojects.

project then can be thought of as being composed of interacting subprojects. In general each phase would correspond to a subproject (Fig. 2.4).

2.3.4 Project life cycles

Together the project phases are referred to as the *project life cycle*. The term cycle is reasonably well accepted terminology amongst project personnel. The choice of the term is regarded by some people as unfortunate because it is argued that there is no cycle involved – a new project doesn't grow following the death of a project by analogy with biological systems. To other people the term cycle is an appropriate term as it is argued that a project starts from nothing, develops a life of its own then passes on to nothing, thus completing a loop. The term life cycle emphasises this time basis of projects. This is particularly evident when viewing the growth and decline of the project team, utilisation of resources etc over the life of a project. The end result of a project of course is generally something tangible that continues after the project is finished.

2.4 SOME HUMOUR

Some humorous phases of a project that may be seen on office notice boards are:

1. INITIAL ENTHUSIASM
2. TOTAL DISILLUSIONMENT
3. BLIND PANIC
4. PASSING THE BUCK
5. BLAMING THE INNOCENT

1. ENTHUSIASM
2. DISENCHANTMENT
3. PANIC
4. SEARCH FOR THE GUILTY
5. PUNISHMENT OF THE INNOCENT
6. DECORATION OF THE NON-PARTICIPANTS
7 ENRICHMENT OF THE LEGAL PROFESSION

1. ENTHUSIASM
2. DISILLUSIONMENT
3. PANIC
4. SEARCH FOR THE GUILTY
5. PUNISHING THE INNOCENT
6. PRAISING THOSE NOT INVOLVED

1. ENTHUSIASM
2. CONFUSION
3. DISILLUSIONMENT
4. SEARCH FOR CULPRITS
5. PUNISHMENT OF THE INNOCENT
6. DECORATION OF NON-PARTICIPANTS

All of the above four examples are very similar. Presumably they started off as one version and became modified as they were communicated from one office to another. All four examples were obtained from different project managers' offices. Some people would say that they are a bit too close to the truth.

Some related cynical musings include the fiction (ASCE, 1993):

Phase 1 – fear: A frazzled engineer slumps in his chair, stares glumly at his profit/loss statement, chomps antacid tablets like candy and worries about where his next job is coming from.

Phase 2 – self-delusion: Rejecting harsh reality, the engineer busies himself with the foolhardy pursuit of illusory projects. Phone calls are made, letters of interest are mailed, and false flattery is lavishly doled out. The irrational hope of this phase can be likened to placing a written message in a bottle, throwing it into the Atlantic Ocean and expecting to hear back from your cousin in England.

Phase 3 – prevarication: A marketing offensive is launched with zealous fervour, complete disregard for the facts and a proposed budget just slightly larger than the profit that could ever be realised from the project being pursued.

Phase 4 – glory: Through some strange twist of fate the project is awarded, backs are slapped, champagne corks are popped and all is right with the universe.

Phase 5 – boredom: Excitement turns to tedium as the staff wallows like overfed sows in an overabundance of chargeable man-hours and a deadline in the distant and unforeseeable future.

Phase 6 – confusion: Where did the file go? Who's the Project Manager? What was it we were supposed to be designing again?

Phase 7 – panic!: The budget is exhausted, the schedule is blown and the major technical issues are unresolved. The stage is set for ...

Phase 8 – finger pointing: A tidal wave of creative energy is unleashed as an imaginatively conceived assortment of vendors, subconsultants, government officials, bad weather, illnesses, and various acts of God are trotted out as scapegoats responsible for the unfortunate status of the project. An engineer's years of schooling are put to full use as this elaborate version of my 'Dog Peed on My Homework Assignment' is developed.

Phase 9 – wonder: A wide-eyed engineer is overwhelmed with childlike amazement as he gazes at the unbelievable result of this irrational, painful, and morally reprehensible process – a successfully completed engineering project.

Phase 10 – fear: A frazzled engineer slumps in his chair, stares glumly at his profit/loss statement, chomps antacid tablets like candy and worries about where his next job is coming from.

Of a non-humorous nature, there are also the simplistic:

1. THINK
2. DO

and

1. START
2. MIDDLE BIT
3. FINISH

A phasing with rhyming appeal, based on a team building analogy, used by some people is:

1. FORMING
2. STORMING
3. NORMING
4. PERFORMING
5. ADJOURNING

There is also the biological life cycle phasing:

1. GERMINATION
2. GROWTH
3. MATURITY
4. DEATH

2.5 PROJECT TYPES

There are many different types of projects and nearly as many ways of categorising projects. One useful categorisation is in terms of the degree of definition of the project end-

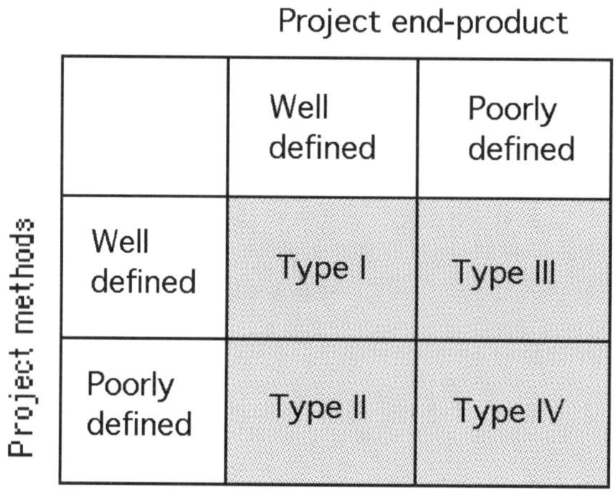

Project end-product

Figure 2.5 A classification of project types (source unknown).

product and the degree of definition of the methods available to achieve the end-product (Fig. 2.5).

Management practices will accordingly differ between the different project types. In Figure 2.5, the different project types have been arbitrarily labelled I, II, III and IV.

Examples of each of these project types are:

Type I projects   – Large engineering projects
Type II projects   – Product development projects, early space projects
Type III projects – Software development projects (the shape of the end-product proceeds with development)
Type IV projects   – Organisational development projects

Particular instances may take these example projects into other categories. It is not intended to say, for example, that all product development projects are of Type II, only that many would be expected to be of Type II.

2.6 PROJECT MANAGEMENT

Project management is the management of these entities called projects discussed above. What management is, is another matter.

2.7 PHASE IMPACT

2.7.1 General

In the early phases of a project, little money is spent but the work done in these phases is believed to have a large impact on the outcome of the project.

It is believed important to get it right in the early project phases as changes down the line are considered to have a bigger impact on the total project cost compared with changes occurring early in the project life.

2.7.2 The influence of early project phase work

Recommended good project management practice is that proper attention be paid to the first phases of a project; they are generally accepted as the most critical phases of a pro-ject. Much is said about the importance and influence of early project phase work on the outcome of a project; such work, low in budget terms, is said to have a large impact on a project's outcome, and in particular its end cost, more than work in any other project phase; without due consideration and adequate effort, project failure is considered almost certain.

The project management literature uses words such as 'vital' and 'crucial' when dis-cussing early project phase work such as feasibility studies, risk analysis, responsibility establishment and requirements definition. There is a widely held belief in the importance of early project phase work.

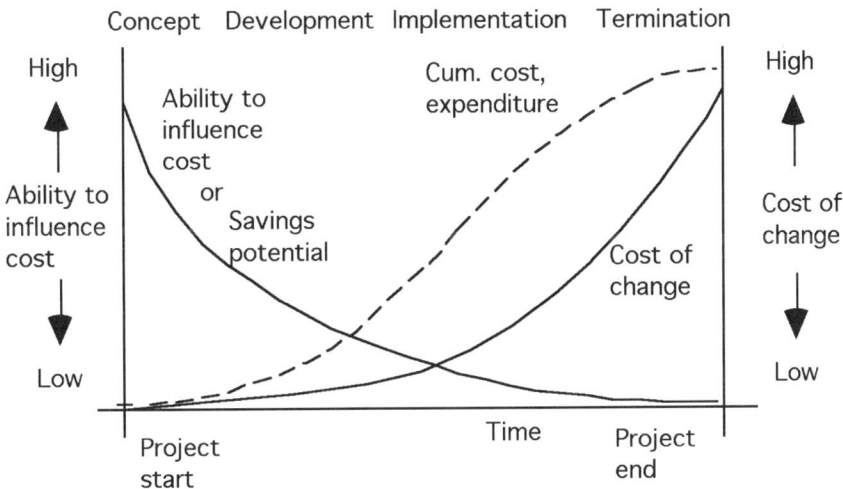

Figure 2.6 Cost influence and changes over a project's duration.

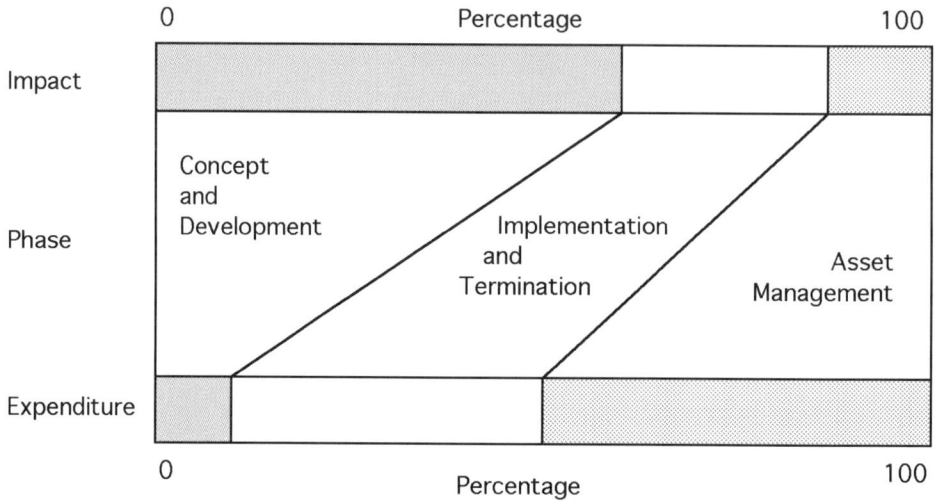

Figure 2.7 Typical expenditure impact of project phases.

Figures such as Figures 2.6 and 2.7 are often quoted. (How such diagrams are obtained, however, is never explained.) The figures indicate that the ability of any decision to influence the outcome of a project is very high at the project's inception when expenditures are relatively low, but reduces with time, while the cost impact of changes grows as the project progresses.

However, it is not possible to demonstrate this in any objective way. (In the terms of Chapter 16, the inverse investigation problem cannot be solved.) There is only experiential observation or anecdotes of projects to support the popular contention. To that could be added 'gut feel', judgement and perhaps logic; it seems a reasonable proposition. It seems reasonable to say that the likelihood of the benefits (cost savings or reduced completion times) of doing early project phase work exceeding their cost is large, and there are no benefits gained and only potential losses by not carrying out the work.

Experience says that preparation and planning in the early phases of a project reduces the chance of a project under-performing and has a significant influence on project cost. Logic says that in order to control project outcomes such as cost, duration and quality, planning for all possible eventualities should be undertaken, and the earlier in the project this occurs, the more closely and effectively the project can be controlled.

Objectivity is not possible because, although many projects are similar, no two projects are the same. A 'control' project cannot be set up, against which other projects with different 'early phase work' are compared; the closest that may be obtained is to observe two similar projects but with different front-ends, but this is inconclusive. Each project cannot be done twice – firstly with preparatory or front-end management work and secondly without. Medical experiments on people, it might be argued suffer the same problem, but get around the issue to a certain extent by conducting trials on large numbers of people. Conducting trials on large numbers of projects, each containing multiple factors that have the ability to influence the projects' outcomes, is not feasible. Individual factors cannot be varied one at a time, while all other factors are kept constant. There is not a one-to-one relationship between any project inputs/factors and project outputs; to a certain extent, a project (as a system) is almost uncontrollable in systems engineering (Kalman *controllability*) terms; altering any one input/factor affects multiple outputs. There are also *observability* (in the sense of Kalman) troubles. The benefits purported to be due to early phase work may be due solely or partly to some other project factors.

Performing one project activity/task affects other activities, and the outcome of other activities. And on large projects there are large numbers of activities, most of which are interdependent. The start of a project represents the point where there is most freedom to change things. As a project progresses, each decision made and activity completed constrains subsequent decisions and activities. It might then be reasoned that the start of a project is the point at which there is the greatest ability to influence project outcomes. However, although reasonable, this can't be proven.

An alternative is to approach the argument from the double negative perspective, namely by reasoning that if early project work is not done, then project outcomes will almost certainly be undesirable. And this is supported by many case projects that have performed badly where early work was not carried out or not carried out properly. The conclusion drawn is that there is value in performing early project phase work. Again this is reasonable, but can't be proven.

2.7.3 Some studies

A number of studies have been reported trying to demonstrate a link between early project phase work and project success. Typically they involve examining multiple real projects, and not any deliberate experimental projects. Projects are categorised, in a pseudo-rigorous fashion, as high, medium or low in terms of early project phase effort and the (falsely drawn) conclusion then follows that there is a positive correlation between effort expended in the early project phases and ultimate project success; the cost and time taken to undertake the early project phase work is reasoned to be outweighed by the resultant improved project performance. The studies give an air of objectivity, but remain largely subjective.

In addition to the logic issues mentioned earlier, there are a number of deficiencies in the studies contributing to the falsely drawn conclusions.

- Effort is not defined, nor its measurement. How are 'high', 'medium' and 'low' defined? Which components of effort are important, and which are optional?
- Terminology and processes are not standardised.
- Communication of data is loose.
- There is no common basis for project evaluation.
- There is no one-to-one relationship between early project phase work and project success.
- The factors that define success and early project phase effort are undefined. Weightings are subjectively chosen.

- Questionnaires are used as a basis for data collection; data tends to be descriptive, rather than quantitative; closed and open-ended questions are used. Data cannot be transferred in a meaningful way.

2.7.4 Reasons for not doing early project phase work

There are a number of reasons/excuses why the early phases of a project are rushed, prematurely completed, or given insufficient attention.

An argument may be made that money cannot be afforded to be spent on conceptual work for projects that may not be approved and may not proceed.

Budgets may be set not allowing for much preliminary work. The case for performing early project phase work may be argued subjectively with reasonable strength, but requires effort greater than arguing the case for later-phase physical, 'bricks-and-mortar' work, particularly in justifying a budget to an owner. Not being able to objectively argue the case for spending money on early project phase work doesn't help; subjective arguments are easily countered by someone holding different opinions. During the later project phases, the owner can actually see something materialising, and see something tangible for its money. Early project phase work includes feasibility studies, alternative generation and thinking, none of which form part of the end-product.

Owners commonly have a poor perception of the value for money (output/money spent; benefits commensurate with, or exceeding the costs) of early project phase work and the non-physical work performed by professionals such as project managers and designers. Many owners are ill-informed buyers. Although the cost of the work done by professionals, such as project managers and designers, is very small and only amounts to a few percent of the project cost, and a fraction of a percent of the life cycle cost of a product, owners still think this is money wasted, and the cost of such work is not fair for the output received. The output of early project phase work is typically documents, and not seen by owners to be worth the money being asked to pay for them. This is exacerbated by the typically higher rates paid to such professionals over those that do physical work; early project phase time is seen as more expensive; higher pay rates are equated by ill-informed owners to lower value for money.

Some consultants overcome the perception of owners that they are not getting value for money from the early project phase work by producing voluminous padded reports, with no more content than a more reasonably-sized report, in order to support their fees. As well, their bids for the work may be submitted unbalanced – the preliminary phase work is bid under cost and the shortfall made up with increased fees for the design and documentation phase. Although not ideal, it provides a practical solution to the problem and still allows the early project phase work to be adequately completed.

Early phase work may not be carried out because of delayed or late commitment of owners to a project, while at the same time not changing the completion date. This compresses the available time to properly carry out the early phase work. When there is no guarantee that the preliminary work will identify time savings that are greater than the time spent on the preliminary work, or cost savings greater than the money spent on the preliminary work, the preliminary work doesn't get done. Fast-tracked projects suffer similarly, where also cost savings may only be secondary to time savings. There may also be a push by owners or a perceived pull from industry and the market to get their product to the market sooner in an attempt to get a competitive advantage. The projects proceed with fingers crossed that nothing unexpected will occur.

Early project phase work may be also downplayed because of the lack of definitive methods for dealing with the associated issues, a problem which pervades most of the practice of management. Attention then is concentrated on the few well established techniques (which usually apply to later phase issues) at the expense of everything else. This is magnified by the lack of education of the people entrusted to do the work. With a lack of understanding of what has to be done in the early project phases, the work is glossed over or disguised with 'smoke and mirrors'. The lack of understanding extends to what is happening in the subsequent project phases, and the relationship between the work in the different phases.

Case example

Conveyor design and construct

The scope provided by the owner was to design and construct a stockpile conveyor to transport ore from the shaft skip of an underground mine to a stockpile area. The construction site for the project was situated on a historical mining lease, where mining had been extensive in the later part of the 19th century but had ceased due to technological limitations in the early years of the 20th century. The major surface infrastructure had been decommissioned and removed from the site as part of the mine closure and the site had since become overgrown with substantial vegetation. With the advent of modern mining techniques and technology the mine was now being redeveloped.

An analysis of the project provides evidence to support the contention that the ability to influence cost was greatest at the start of the project, an opportunity which, in this case, was lost due to poor early preparation.

Early phase work included planning and preparation relating to site investigation, geotechnical investigation, estimation/budget, design issues, schedule, resourcing, quality, and finance. These activities were not carried out in sufficient detail when performing the cost estimate for the project and the repercussions incurred are detailed below.

Site investigation and survey

A detailed site investigation and survey was not conducted resulting in the failure to identify existing footings and pedestals from the demolished winder houses of the original mine infrastructure. The mine site was heritage listed and the footings were not able to be removed. Their presence was not detected until the construction phase, and their preservation and the work included in doing so, was not allowed for in the original budget or schedule. This delayed the construction schedule leading to adverse variations in cost.

Geotechnical investigation

The geotechnical investigation was done in parallel with the design, with the geotechnical report not being available until two thirds of the way through the design period. The design was progressed on assumptions of the underlying ground condition. The geotechnical report revealed very poor sub-base in the position of a major trestle support for the structure. Further investigation of historical mine records revealed the presence of one of the main shafts (of the previous mining operation) that had been collapsed and filled to only some metres below ground level.

To avoid this unstable ground, the line of the conveyor had to be moved, resulting in major rework to the site layout, civil and mechanical drawings. Drafting hours and cost increased, and delays to the schedule were incurred.

Resourcing

The original budget had not considered resourcing for the project in detail. A single engineer was provided to perform the project management, design, design office supervision, contract development, procurement and construction management. This proved severely inadequate resulting in the mobilisation of additional people to the design, adversely affecting the project cost.

Budgetary controls

An official control budget was never established for the job. The project was launched straight into the design phase without consolidating the information prepared in the estimate. Hence, tracking costs proved difficult because there was no fixed budget to compare the costs against. This made project reporting to the owner difficult and time consuming. It also had the severe adverse effect of reducing owner confidence in the project manager. With no hours allowed for budget creation, this problem was never rectified.

Due consideration of the complexity of the required structure

The conveyor design incorporated a main transport conveyor and a secondary reversing conveyor at the head end to transport ore to the stockpile or mullock to the waste stockpile. The complexity of the head end structure required to support the reversing conveyor was not considered in the cost estimation. This resulted in overruns in design and drafting hours, leading to overruns on the original project schedule. Drafting hours and cost increased. Additional steelwork in the head truss led to a variance on fabrication costs.

Consideration of the braking system for the reversing conveyor

Being an inclined reversing conveyor, a brake was required to stop the belt rolling backwards when the belt was not operational. The braking system was not considered during the initial costing estimate. During the procurement stage of the project, the delivery period for the brake system (available only from overseas) was substantially longer than the original scheduled construction period for the project. The delays that would have been expected from this oversight were fortunately absorbed within owner instigated delays.

Quality supervision in the fabricator's work

Allowance was not made for quality supervision in the fabricator's work. Minimal periodic supervision was carried out by the owner using staff untrained in the field of structural fabrication. Paint thickness testing conducted by the construction manager during construction revealed shortcomings in the application of the protective coating system, and this had to be rectified in-situ.

Non-destructive testing of critical structural welds called for in the contract specification was not carried out by the fabricator. This error only became apparent during the commissioning stage of the project, resulting in a delay in the commissioning of the system. This resulted in a delay in underground mining causing significant loss to the owner.

Project financing

Project financing was the responsibility of the owner. It became evident only toward the end of the procurement phase of the project, when orders required placement, that the owner had not progressed financing of the project, leading to the inability to place orders for site work and equipment. This had a significant impact on the original project schedule, delaying the order placement for the major contracts by one month. Secondly, finance was only initially raised for the civil work and structural fabrication. Financing was not raised for the installation of the structure until after the completion of the fabrication. This incurred variations from the fabricator for delayed site establishment and off-site storage of the structure.

The problems encountered during the above mentioned project were due mainly to poor and insufficient preparation and planning in the early phases of the project. The project effectively commenced at the design stage and, due to insufficient budget and the constraints of a fast-track schedule, the ability to positively affect the cost of the project was severely handicapped.

Given the same set of circumstances, the problems encountered in this project could well be repeated. Conversely, given adequate preparation and planning in the early phase work the problems encountered could readily have been avoided and the cost of the project significantly reduced. This is an example of a project where initial work in the early phases of the project could have identified potential areas for concern; allowance could have been made for those manageable risks and a contingency included for those risks that were not considered manageable.

Case example

Winery upgrades

(i) This case study discusses the involvement of a consultant in the upgrade of wineries. Typically this involvement in the projects occurs after the owner has determined that the project is feasible, has done a preliminary cost estimate in-house and has allocated funds to the project. The consultant's initial task is to meet with the owner's chief engineer and the relevant winery manager and determine the exact scope (of work) for the project and prepare a detailed capital cost estimate for the work. The consultant prepares drawings for various layout options and cost estimates for each option. The consultant also prepares programs for each option, showing when the owner needs to make various decisions in order to meet a required end date, which is usually prior to the next vintage.

In this way the consultant is able to show the winery manager how s/he can obtain the most items from a 'wish list' given the budget constraint of the project. This process also forces the winery manager to prioritise his/her requirements. In addition, it flags to the chief engineer when certain decisions (approval to proceed with design and documentation, ordering of specific winemaking equipment etc) need to be made in order to meet the required end date.

By putting a dollar and time figure on the impacts of each option to allow the owner to choose the option which gives it the maximum return for its given resources, the consultant demonstrates to the owner the value in the early phase work. By involving the owner in the process and showing it the various layout, cost and program options considered, the consultant is also able to demonstrate the amount of background work that goes into the preliminary stages and the owner is therefore happier that it is getting value for its money. (Although it is necessary for the owner to be involved in the decision making process between the various options in any case, the consultant makes a point of ensuring the owner sees the work that is put into this phase of the project.)

(ii) Master planning work was done for an old winery, that had had a number of owners in its lifetime. Its development had been on an ad hoc basis, with little thought being made to the relationship between the various developments or to future requirements.

The master planning work identified the crushing capacity and hence process equipment requirements, for the winery for the next ten years. Various layout options were prepared showing how the winery could achieve these, and the advantages and disadvantages of each option were identified. Based on this, the owner was able to pick the features of each option that it liked, and a final plan and a preliminary cost estimate were prepared. The owner at first objected to the cost of this work. However, the results of this master planning work were directly used in planning the layout for a future project, both in locating new equipment being provided under the current project and in identifying areas that had to be left free for future development. The consultant was hence able to demonstrate the value of the work.

(iii) In both of the above examples the value of the early project phase work was demonstrated to the owner only after it had been completed. To get the owners' approval to do the work in the first place required arguments that were subjective – 'this is what this work achieved on previous projects for other owners, and we believe it can do the same for you.' However, the owners can now see for themselves the value in the work and it is now carried out as a matter of course.

Case example

Land development

A developer acquired land to build a multi-storey office block, residential apartments and a shopping centre. As part of the existing environmental (natural) laws, the developer was required to get an environmental (natural) audit done on the site. The developer approached an environmental (natural) consultancy to do the investigation to get the audit approved.

The consulting company conducted a brief site history and found the site had been used for industrial purposes. From this it concluded that a certain number of test sites would be required for sampling to get the audit approved. The consultant provided the owner with a proposed scope (of work) and a lump sum cost for the work to be done; it was thought it would take about one month to complete

One week into the investigation it was realised the investigation would be totally inadequate because the site geology and previous industrial uses had allowed the site to become heavily contaminated. The initial investigation took two months and eventually cost over twice as much as originally envisaged. This unexpected outcome did not please the owner because it assumed the site would be fine and would pass the audit straightaway.

A second investigation of the site was approved to delineate the areas of worst contamination. Prior to putting in a cost for the investigation, some of the pre-proposal work undertaken was:

• Assigning an experienced project team to the investigation, with the responsibility and requirements of each team member clearly defined.
• A detailed site history of the site, traced as far back as possible to identify land uses and possible contaminants.
• Liaison with the owner to indicate the potential problems with the site and ensure that the project was still feasible.
• Discussions with the environmental (natural) auditor to determine a scope for the investigation.
• Determination of the timing of each activity.
• Identification of any unforeseen events and the development of procedures to manage them for both the consultant and the owner.
• A cost estimate of the project based on rates.

The second investigation was approved by the owner and was undertaken by the consulting company. The outcome of the investigation was much more favourable than the first for the consultancy with a small profit being made by the consulting company and the investigation being completed on time.

However the most important outcome of the second investigation was the improved relationship with the owner. Although the site was still heavily contaminated and a lot of money would have to be spent on redemption, the owner knew the risks because it was kept informed at all times and was then able to develop contingencies for them as well as keeping its investors informed.

The two projects, of similar nature, conducted on the same site had two different outcomes. The second project had proper front-end management. But was conducted with the benefit of hindsight from the first project. It was believed by all involved that the difference between the two projects highlighted the importance that proper front-end management has on a project.

EXERCISES

1. Based on projects with which you have been associated, write down your definition of a project.

Does a definition of a project have to make reference to such matters as activities, resources and planning? Justify your position.

2. Some typical examples of projects include:
• Launching a new venture.
• Developing a new product.
• Effecting a change in structure, staffing, system or style in an existing organisation.
• Turning a poor performance situation into a satisfactory one within a target period.
• Designing and producing a new transportation vehicle.
• Designing and constructing a building or facility.
• Implementing an urban or rural development program.

What are the attributes of these projects? What is it that makes them unique?

3. A project presumably starts with an idea or a dream. Would you include the activity of having an idea or a dream in the first project phase or does it come before the first project phase? Is the idea or dream part of the project or not?

4. Based on projects with which you have been involved, identify any phases that the projects went through.

For any of your projects, were the activities sufficiently blurred that no phasing was distinguishable?

Is there a need to identify a project's phases from either a theoretical or practical viewpoint? Give your reasons.

5. Projects commonly involve feedback of information. For example, although in the initial stages of a project a lot of time may go into planning the project, the plan may be a fluid document that may change slightly as new information comes to hand during the project. Is it correct then to allocate the planning activity to one project phase?

Other activities of a project manager may occur in all project phases. For example negotiating skills are used throughout a project. To which project phase then do you allocate negotiation?

To extend this reasoning, what criteria do you use in defining the project phases?

6. Think – Do. For the simplest of tasks, people commonly talk in terms of a *thinking phase* and an *action phase* – the planning and the execution. How useful is such an approach to a project (a collection of such tasks), or is it just an approach that stresses the need for thinking before doing, and also stresses the balance needed between planning and execution?

Germination – Growth – Maturity – Death. Is an analogy with biological life cycle phasing of use in projects?

7. Give examples (other than those above) of the different project types – Types I, II, III, and IV.

8. Comment on how you believe project management practices would differ for each project type – Types I, II, III, and IV.

9. How does the treatment of risk change between the different project types (Types I, II, III, and IV)? Are certain project types 'riskier' than others? Consider with respect to meeting cost, time and quality targets.

CHAPTER 3

What is Project Management?

3.1 INTRODUCTION

Perhaps the most difficult question that project managers are asked is 'what is project management?' There is an associated difficulty of defining 'management' even by most people who perform management duties in their daily lives; this is of course complicated by many people adopting the perceived prestigious title '... manager' when a more appropriate title is applicable.

Defining project management has been remarked as being much like the dilemma in Alice in Wonderland:

> 'When I use a word,' Humpty Dumpty said in a rather scornful tone, 'it means just what I choose it to mean, neither more nor less'.
> 'The question is', said Alice, 'whether you can make words mean so many different things'.
>
> ('Through the Looking Glass', Ch. 6, Lewis Carroll)

Many people are able to recognise project management (PM) when they see it but find it harder to express what it is in words. Others have a coloured view. This is reminiscent of the fable of Figure 3.1.

3.2 DEFINITIONS

Many published definitions of project management are partly circular describing management as management. Two possible (but not entirely satisfactory) definitions are:
- *Project management is the overall planning, control and coordination of a project, from inception through to completion. It is the process by which the responsibilities for all phases of a project are combined within one multi-disciplinary function.*
- *Project management is the management of change.*

The notion of a multidisciplinary function focus in a definition is appealing. The use of the terms planning, control and coordination come from one of the accepted ways of describing the functions of management; in addition, in some management texts you may see further functions mentioned such as organising, staffing and directing.

It was six men of Indostan
 To learning much inclined
Who went to see the Elephant
 (Though all of them were blind)
That each by observation
 Might satisfy his mind.

The First approached the Elephant
 And happening to fall
Against his broad and sturdy side
 At once began to bawl:
'God bless me! but the Elephant
 Is very like a wall!'

The Second, feeling of the tusk,
 Cried, 'Ho! what have we here
So very round and smooth and sharp
 To me 'tis mighty clear
This wonder of an Elephant
 Is very like a spear!'

The Third approached the animal
 And happening to take
The squirming trunk within his hands,
 Thus boldly up and spake:
'I see', quoth he, 'the Elephant
 Is very like a snake!'

The Fourth reached out an eager hand,
 And felt about the knee
'What most this wondrous beast is like,
 Is mighty plain', quoth he;
' 'Tis clear enough the Elephant
 Is very like a tree!

The Fifth who chanced to touch the ear
 Said: 'E'en the blindest man
Can tell what this resembles most;
 Deny the fact who can,
This marvel of an Elephant
 Is very like a fan!'

The Sixth no sooner had begun
 About the beast to grope,
Than, seizing on the swinging tail,
 That fell within his scope,
'I see', quoth he, 'the Elephant
 Is very like a rope!'

And so these men of Indostan
 Disputed loud and long,
Each in his own opinion,
 Exceeding stiff and strong,
Though each was partly in the right,
 And all were in the wrong!

The Moral:

So oft in theologic wars,
 The disputants, I ween,
Rail on in utter ignorance,
 Of what each other mean,
And prate about an Elephant
 Not one of them has seen!

John Godfrey Saxe (1816-1887

Figure 3.1 The Blind Men and the Elephant (after Bowen, 1987).

The second definition tries to take into account the dynamic nature of projects and in particular the transition of projects from one phase to another. Activities, staff resources and circumstances are continually changing or evolving throughout the life of a project. Management styles based on change reporting find favour with some people. This is to be compared with the management of an ongoing business where tomorrow's activities, staff and resources are similar to today's. Ongoing businesses tend to be quasi-static by comparison with projects. The degree of dynamism of projects may vary of course depending on the lifespan of a project; some project lifespans are measured in hours, some in years.

Effective project management encompasses many skills including repeated logical decision making in an uncertain environment, commonsense and perception. This is carried out within a framework of a knowledge of management principles and a systematic way of organising people, documentation, finances and commercial practices.

3.3 WAYS OF LOOKING AT PROJECT MANAGEMENT

3.3.1 What a project manager does

By the end of this book it is hoped that the reader has a good feel for what project management is even if a definitive definition is not forthcoming. Most definitions are usually in terms of what a project manager does. They often come from studies made of actual projects and observations carried out on the project managers. There are three favoured ways of describing what a project manager does, but other ways do exist:

(i) A 'classical' management approach

Management is commonly described in general management texts in terms of:
• Planning.
• Organising.
• Staffing.
• Directing.
• Controlling.
• Coordinating.

Some people find such a breakdown useful. Others find such a breakdown so unsystematic as to be of no use. It represents cutting the cake in X different ways at the same time, and consequently is not internally consistent.

(ii) A management function approach

An alternative view of project management is to regard it as a collection or integration of subfunctions:
• Scope management.
• Quality management.
• Time management.
• Cost management.
• Risk management.
• Contract/procurement management.

- Human resources management.
- Communication management.

The origins of such a breakdown are unclear. It represents cutting the cake in Y different ways at the same time, and consequently is not internally consistent. Many functions seem to be there based on the 'string on the finger principle', that is if the string is not on the finger, the person will forget about it. The function of quality management could be argued has only a small body of knowledge of its own and represents a redundancy of information. The function of risk management does no more than look at the stochastic version of the problem covered in other functions and hence does not warrant its own special treatment. The choice of the term 'time management' is unfortunate. Inert resources (materials, equipment, plant, ...) are not mentioned.

(iii) Chronological approach

A project may be seen as going through a number of phases. There is no consensus as to the naming of phases or to the number of phases. In terms of management, many people feel comfortable with a chronological way of describing a project and the activities involved.

3.3.2 General

Each of the three favoured ways of describing what a project manager does has its own followers. They are different ways of slicing the same cake. The second way has advantages over the first way in terms of communicating with the owner and for progress reporting and documentation purposes. It may also fit in well with functional/departmental groupings in an existing organisation. The third way aligns more with the way people actually think about projects. Contracts for project management services could be written for all three ways.

  The divisions of management functions as suggested above are only some of many possible divisions. They may provide a useful organisation of knowledge although they may not be to everyone's liking.

  Both the classical management and the management function approaches represent, in evolutionary terms, going out on a branch. They were possibly useful in the short term for practitioners, but have hi-jacked fundamental thought on project management and the development of a rational framework for management. They represent dead ends, useful for a while, but not fundamental and robust enough to promote development. They will become extinct. It is time to get back to the evolutionary tree trunk, such that beneficial development can proceed.

3.4 IS PROJECT MANAGEMENT DIFFERENT TO GENERAL MANAGEMENT?

3.4.1 Comparison

To many people, the central issues of project management are related to time and change. Projects have definite start and finish dates. In contrast, managers of ongoing enterprises may look to long-term goals and the long-term success of those enterprises and invoke administrative and management practices and technology to suit.

This necessarily implies different management skills. The human resources management aspects can be quite different particularly in regard to matters of organising and leading.

Organisational structures for projects differ from those in ongoing enterprises. The situation becomes more involved when an ongoing enterprise additionally takes on project work; here matrix organisational structures may be used leading to divided responsibilities and authorities. Authority relationships in ongoing enterprises are seen as lesser in importance than in projects.

Compared to managing a project, with its uniqueness, managing an established on-going enterprise could expect to be characterised by:
• More certainty over time.
• Tasks may be repetitive.
• Well understood roles and interpersonal relationships.
• Relatively stable work situation.

Some people take issue with the view that managing an on-going enterprise is different to managing a project, pointing out that on-going enterprises and firms, in today's world, are continually subject to change. Change is brought about by two causes:
• Externally forced through technology advances, competitors, regulations, economic conditions and so on.
• Internally applied, on the assumption the present situation can be improved or needs improving. Change is introduced even when there is no indication that a change will bring about improved benefits, or the cost of the change is justified. There is a thriving population of management consultants and a thriving trade in the popular management literature all imploring change as a placebo; there rationale is not much more than 'a change is as good as a holiday'.

Change management has become another buzz word.

The distinction between projects and on-going enterprises, however, that is referred to here is the rate of change and the dynamics of change.

3.4.2 Change

An aspect of project management, on a typical major project, that is decidedly different from general management is the continually changing nature of the project organisation as the project progresses through its various phases. This requires a project manager to be continually reviewing and adjusting the organisational structure of the project team to ensure the effective use of resources on the project.

This aspect of continual change can be one of the most exciting challenges and for many people one of the most difficult to manage. The changes require managing, and a project requires participants who are capable of adapting to the changes. Depending on the nature of the project and its duration, the changes can be seen as gradual evolutionary changes or radical structural rearrangements. For example, on a typical major construction project the gradual changes can be seen in the build up and decline of the detailed engineering design team whist a radical organisation change occurs with the completion of the design and the commencement of construction. This latter radical change can see a wholesale change in the actual participants in the project and a change in the importance of their part in the project as time progresses.

A stimulus to project participants is to see the completed facility at the end of the project and to identify with having contributed to its realisation.

3.4.3 Management by projects

For some work practices it is possible to think of them as projects and apply project management principles. The term 'management by projects' is sometimes used to describe such an approach, and is sometimes abbreviated to 'MBP'.

Many people carry out this practice without adopting a formal term to describe the process or recognising that in fact this is what is happening.

Work can be broken into parcels and each of these parcels regarded as a project with its own objectives and constraints.

The approach provides a structured and systematic way of solving general management problems. On identification of a work parcel as a project, attention then shifts to scope definition, identification of necessary resources etc. This focuses attention on the particular work parcel and provides something recognisable that the employees can take responsibility for.

In terms of a business entity with quarterly, yearly and so on targets, there is also the possibility of managing this by project management methods. Uncertainties in the business scenario are treated in a similar fashion to uncertainties in a project scenario.

It is a similar approach to where projects themselves are regarded as being made up of a number of tasks and each of these tasks is treated as a project in its own right. For example, the preparation of a feasibility report, although part of a larger project, for the purposes of undertaking this task, may be regarded as a project, albeit with special characteristics.

3.4.4 Projects and general management

Ideas from general management have not been totally embraced by the project community. The reasons for this are unclear. It may be that project personnel regard general management theories as inapplicable. As well, certain industries are inward looking and are unaware of developments and practices outside those industries.

Much of the voluminous paperback publications in the general management area are regarded by technically-oriented project people as shallow, inexact and opinion- and anecdote-oriented rather than being objective. They are likened to jumping up and down on the one spot and sometimes even going backwards rather than advancing the state-of-the-art. This has the effect of alienating project personnel from readings in the general management area.

The situation is not helped by the plethora of best-selling paperbacks, videotapes and audio cassette tapes with eye-catching titles based on peoples' successful (Western and Eastern) business practices but which cannot be generalised beyond those peoples' peculiar circumstances and environments. Such books do get read by a few project personnel but more from a recreational reading viewpoint. Few rigorous management texts seem to find their way onto the bookshelves of project personnel.

Generally, the state-of-the-art of project management and general management may both be improved by looking beyond their own bounds to the writings and practices of the other management form.

3.5 PROJECT MANAGEMENT AS A DISCIPLINE

The reputation of project management suffers from people undertaking the task without being aware of all that is necessary to perform the task properly. People without the necessary skills and qualifications practising project management damage the discipline. At present anyone can hang up a shingle or advertise that s/he is a project manager. Inevitably some owners are duped into hiring people without the appropriate skills. Accreditation of project managers is one suggested way of rectifying this situation.

There are a number of professional organisations worldwide set up to advance the profession of project management. Attention is directed, by these professional organisations, to the development of the state-of-the-art of project management, educational programs, accreditation and professionalism, and to the establishment of an identifiable body of knowledge for project management (in one case referred to as PMBOK, the Project Management Body of Knowledge).

Project management, while developing its own approach and techniques, borrows from knowledge available in general management as well as the disciplines in which it operates. There now exists a *generic* body of fundamental knowledge in project management, applicable to all applications, together with approaches and techniques *specific* to particular industries or technologies.

EXERCISES

1. Which of the above three ways of describing what a project manager does appeals to you? State your reasons.

Try to relate the three ways.

2. In many ways the field of project management is very crude when compared, for example, with established engineering disciplines of age where there is little chance for

you to contribute anything meaningful. However there is potential for someone to make a significant contribution to the state-of-the-art of thinking about project management. As you read further through the book keep this matter in mind.

Project management and general management are very young by comparison to established engineering disciplines.

Most of the contributors to general management seem to have come from non-technical backgrounds. The important contributions to project management seem to have come from people with technical backgrounds.

This has led to general management models being primarily verbal models. However models preferred by people with technical backgrounds are quantitative models and often mathematical in form. The imprecision of verbal models to a person with a technical background is always a concern.

In the distant future, management models will be on a par with engineering models. The argument advanced against this is that management is concerned with behaviour, and people are extremely fuzzy entities. The counter to this is that perhaps we are not looking at the modelling process in the right way – some lateral thinking is needed.

3. Standards (of materials, finish, workmanship, ...), time and cost are interrelated and cannot be considered as isolated entities. A triangle is sometimes used to pictorially show the relationships in a qualitative way, either:

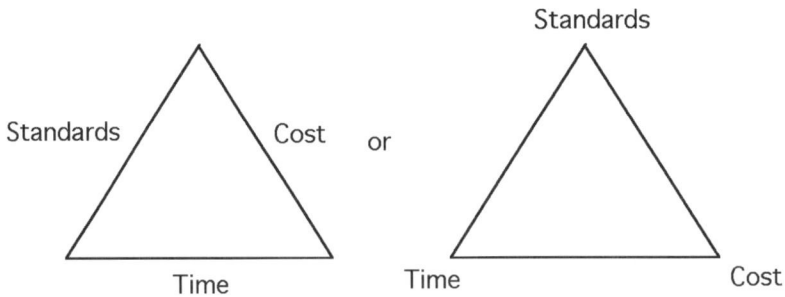

Consider a project with which you have been associated. For the important stakeholders in that project identify where in the triangle you think their priorities were, and how this point moved as the project progressed.

4. Various sources give different chronological listings of activities undertaken by project managers. Why do you believe there is no consensus or definitive description of project management activities?

5. Which of the above listed differences between an established on-going enterprise and a project do you disagree? Give your reasons. Are there other differences that you can add?

Is the multi-disciplinary characteristic a relevant difference?

6. Compare the mass production of motor cars with the production of a one-off specialist vehicle. Characterise the management differences between these two.

Would it be true to say that in a mass production situation, the people and plant would be specialised, while in a one-off production situation, the people and plant would be required to demonstrate multiple skills/uses?

7. One inference from the term, 'management by projects', is that any person who is a good project manager will automatically be able to be a good general manager. Explain your views on whether you think the inference is credible.

A second inference from the approach is that the body of knowledge of general management can gain from the body of knowledge of project management. Again explain your views on the credibility of this second inference.

8. Consider an assembly line process (mass production) where at each stage of the line a new component is added until at the end of the assembly line a completed item results. This process is repeated continuously in the making of multiple items.

Can you define this process or part of the process as a project?

Is there any advantage in this?

Is it better to devise a management style to suit the process or reformulate the process to fit an existing management style such as project management?

9. Some people intuitively redefine their work into project parcels. Think about your work methods. Are you one of these people?

10. Consider a service industry such as a travel agency, library or a shop. Or consider the routine work of a desk-bound paper shuffler. How could the activities involved be broken into projects?

11. One reason for accreditation is so that the public, and in particular the owners who use project management services, will be able to distinguish between a 'charlatan' and a 'professional' project manager.

There is an argument that accreditation is not necessary; it is the owners who should be educated to recognise proper project management services, and let conventional market forces weed out those without the requisite performance qualifications.

Give your views on the need for accreditation.

Do you see, in the future, that with major owners, only those project managers with accreditation will get work?

How do owners, particularly owners in an industry for a short term, distinguish between a good project manager and someone else without accreditation?

Will owners who have no interest in owning and operating a facility after it is constructed always employ the cheapest project manager irrespective of the project manager's skills and any accreditation process?

12. What differences would you expect between project management as practised in different countries? How would different work habits, salaries, industrial relations structures, cultures and other matters associated with people affect the practice of project management in different countries?

Examine project management books from different countries. Do you observe any differences in practice?

Professional practice in Australia in most fields borrows freely from the rest of the world and also contributes to the state-of-the-art worldwide. Ideas are adjusted to suit local customs. Under such circumstances would you expect project management as practised in Australia to differ significantly from that practised in other English speaking countries?

Is there such a thing as appropriate technology applied to project management when practising, say, in a developed country versus a developing country?

13. Compare management of an ongoing enterprise and project management in terms of:
• Interpersonal roles.
• Informational roles.
• Decisional roles.

14. Compare a manager of an ongoing enterprise and a project manager in terms of the requirement for:
• Technical skills.
• Human skills.
• Conceptual skills.

15. Ideas from general management have not been totally embraced by the project community. What do you think are the reasons for this? Is it that the theories are inapplicable to the project arena? Is it that the project community is unaware of any developments outside its own world? Or is it because of some other reason?

16. What place do you see in project management for general management's mechanistic/rational thinking, human/social thinking or integrative thinking?

17. You are given above some definitions of project management. Can you give other definitions? Do you have a preferred definition?

18. What is common between general management practice and project management practice?

19. Consider: The need for handling people is common between general management and project management, but is the way that they are handled different? What influence do the different organisational structures between an ongoing enterprise and a project have on the way people are managed? Do motivational aspects differ between an ongoing enterprise and a project?

20. Are project people too busy to take the time and stop and look at what they are doing and whether it can be done in a more efficient or effective way? Is it bad logic to say you are too busy to look at ways to do your job better? Is this a similar situation to where some project people start the implementation phase of their project without doing the necessary ground work in the concept and development phases?

Is management situation specific? Can it be expressed in an all-generic form, or is the present state-of-the-art of management a combination of generalisations and situation specific practices?

Is much management practice of the form: this is the way we did it yesterday (or on the last project) and it seemed to work, so this is the we will do it today (or on this project)?

What value is there in having someone detached from your project looking at, and commenting on, your management practices? Why is it that this often only happens after something major has gone wrong, much like a repair-on-breakdown philosophy, instead of a preventative maintenance philosophy?

'Management by objectives' (MBO) involves staff, rather than being told how to carry out their work, being able to carry out their work in any reasonable way as long as they reach clearly defined targets. Progress is periodically reviewed and management assists wherever possible. The approach allows staff to take initiatives and accept responsibilities. In the past, it was a trendy management approach. What is its applicability to projects? [Note, the usage of the term 'objective' is not that used in systems theory.]

Project-based industries, such as construction, are notorious for the small amount of money spent on research devoted to improving management practices and management knowledge. (Note there is a large amount spent on non-Pareto items such as materials, fixings and proprietary items.) Why is the concern for the short-term profit instead of the long-term development of the industry? Why is there so much lack of voluntary sharing of information within the industry? Why does today's profit line drive all activities?

Systems theory or systems engineering provides a rational way of looking at most management matters yet it appears to be ignored by most management training schools? Why is it largely ignored? Is systems' non-verbal basis off-putting to management teachers who, by and large, adopt verbal models for most matters.

What place does simulation have in advancing understanding in management?

21. Project management approaches. Project management can be presented in a number of ways. The most popular are in terms of:
- A chronological approach.
- A management function approach.
- A classical management approach.

Each represents a different way of slicing the cake.

The chronological approach traces the development of a project from the concept phase through to completion. The management function approach looks at project management in terms of management functions such as quality management, scope management, time management, cost management and so on. The classical management approach also uses management functions but described in terms of planning, directing, controlling and so on.

Consider a project with which you have been involved. Subdivide it in the following three ways:
(i)   In terms of planning, controlling, coordinating, organising, staffing and directing functions.
(ii)   In terms of the management of scope, cost, time, quality, risk, contract/procurement, human resources and communication functions.

(iii) In chronological terms according to project phases.

Then carry out the following:
(a) Comment on the suitability or otherwise of each of these three perspectives in helping you think about the management aspects of the project.
(b) Critically appraise the approaches and methods used on your project in terms of the three perspectives.
(c) Do some lateral thinking and try to relate the three perspectives, because after all they are nothing but different ways of describing the same thing.

22. Generic and specific project management. Project management tools and skills have evolved through a number of industries and disciplines. Each discipline brings with it something which is specific to that discipline as well as something that has wider applicability – it is generic in form.

From the body of knowledge on project management as identified by this book, texts and other documents that you have access to, identify (in broad terms) that which could be truly regarded as generic project management knowledge and that which is specific to particular industries or disciplines.

Justify your classification.

Where do you see project management knowledge going from here? What deficiencies exist in our current knowledge?

23. 'Management by objectives'/management by projects. The term 'management by objectives' (MBO) has been around for several decades now. It refers to an approach that has been practised in different disguises for much longer but lacked public appeal because it didn't have a trendy title. The essence of the approach is as follows. Staff, rather than being told how to carry out their work, are able to carry out their work in any reasonable way as long as they reach clearly defined targets. Progress is periodically reviewed and management assists wherever possible. The approach allows staff to take initiatives and accept responsibilities.

Targets alternatively might be called goals, key result areas etc. Note, the usage of the term 'objective' is not that used in systems theory.

Consider, now, project management (PM) or, in non-project situations, management by projects (MBP). The term 'project management' has also been around for several decades now. It too refers to an approach that has been practised in different disguises for much longer but lacked public appeal because it didn't have a trendy title.

In project management, clearly defined project targets may be set. Typically these targets refer to cost, time and quality issues, but not exclusively. Projects are broken down into subprojects and even smaller for manageability reasons; each of these subprojects has targets which flow from the overall project targets. An hierarchical project team is established with work division and specialisation, in a parallel fashion to the project breakdown or work breakdown.

Is the essence of 'management by objectives' different to the essence of project management? Discuss.

CHAPTER 4

Systematic General Problem Solving

4.1 INTRODUCTION

Management involves solving problems. A systematic approach rather than an ad hoc approach could be expected to reap better rewards.

Value analysis (value engineering, value management), constructability and *buildability* studies, *risk management, work study, reengineering* and similar are all shown to be particular versions of the systematic problem solving methodology advanced here.

4.2 THE PROBLEM SOLVING PROCESS

Based on observations, the problem solving process may be broken down to a number of operational steps, functions or phases (Hall, 1962):
• Defining the problem (including issues and situation).
• Selecting objectives.
• Generating ideas, alternatives.
• Analysing ideas, alternatives.
• Selecting the best alternatives (evaluation).
• Action.

As a systems synthesis problem (solved via analysis) or a systems engineering problem, the steps are the same but called:
• Defining the problem (including issues and situation).
• Selecting objectives.
• Synthesising systems.
• Analysing systems.
• Selecting the best alternatives (evaluation).
• Action.

Problem solving and systems engineering are equivalent because the definition of system is sufficiently broad to cover any entity considered in problem solving.

The relationship between the steps is important. It may be found that many steps overlap. For example, in developing one step, the problem solver may have in his/her mind information about other steps which is being concurrently progressed.

Terminology

Synthesis and analysis are used here in the sense defined by the fundamental systems problems of synthesis and analysis (Chapter 16). Synthesis here extends the lay dictionary meaning of combining parts or elements into a whole, to include other ways of system formation. Analysis here extends the lay dictionary meaning of separating the whole into its constituent parts or elements, to include evaluation or deducing consequences.

Note, the fundamental systems problem of systems synthesis may be approached in an iterative analysis fashion, as essentially given here, or directly through optimisation approaches.

Definition

A problem in a general sense might be defined as:
• A situation in need of a change.
• Anything, matter, person etc that is required to be dealt with, solved or overcome.
• Something involving choices of actions.

As well there is the sense of a problem as occurs, for example, in mathematical textbooks of:
• A puzzle, question etc posed for solution or discussion.

With this second sense, all the problem definition and outline has been done by another person. The solution process could accordingly be expected to be shorter.

The first sense is used in the following. The solution to a problem is equated to decisions/actions/controls taken in a synthesis problem. Problem solving methodology is thus no different to design methodology as popularised in the so-called systems engineering approach. See, for example, Hall (1962).

4.3 PROBLEM DEFINITION

The problem definition step transforms an indeterminate situation.

It involves a number of functions including, in various combinations:
• Observation.
• Surveys, data gathering.
• Research.
• Characterising the situation, the issues.
• Examining the environment.
• Understanding the problem, awareness.
• Removing uncertainty, doubt, confusion.
• Hypothesis testing.
• Concept testing.

At the beginning of the step, a person has a vague notion of the problem and in particular vague objectives, and is unsure how to proceed to a solution.

To some people, this step involves an acknowledgment that a problem exists, perhaps admitting this for the first time.

For groups, a common problem has to be recognised and stated.

The step implies a recognition that some inquiry is necessary, even though possible outcomes are unsure. It then develops into a formulation or definition.

Many people play little attention to the early steps of problem solving, or conveniently bypass them because they are considered too difficult. There is often a rush to come up with an answer. Quick decision makers are admired. People who think long and hard about a problem are criticised. (Procrastinators and like people are also criticised but they are deserving of this criticism.) 'Give me your solutions, not your problems.' 'Let's move on and solve this problem.' Yet it is the preparatory steps that are commonly the most important in any endeavour.

In all matters, success depends on preparation. Without preparation there will always be failure.   Confucius

If you don't know where you are going, you will end up somewhere else.
                                                                                          L. Peter

Some writers suggest spending up to half of the total problem solving time on defining the problem.

For textbook 'problems' as presented by teachers, the problem definition step work has been done by the teacher for the reader. The problem is stated by the teacher who has full knowledge of the answer. If the problem solution or answer is known, the problem statement is straightforward. In general it is a 1:many relation (that is, *non unique*) going from the answer to the question, while a 1:1 relation in going from a well-posed question to the answer.

The saying:

A problem well stated is a problem half solved.
                                                        Charles Kettering (1876-1958)

or

A problem well put is half-solved.
                                                                                          Dewey

encapsulates the problem definition step. It attempts to find out what the problem is. It follows that without a defined problem the ensuing steps have no meaning and would be pointless.

There can be a tendency for this step to be shortened in order that a person can 'get into it'. However, it is argued that by building a strong foundation in this step, the later steps progress more effectively. Solutions may start to appear but should be put on hold until the step is fully explored, that is the problem fully defined.

One of the great myths about problem solving is that the most important part of the process is coming up with ideas. Too many people believe this and, as a result, are ineffective problem solvers. Many creative idea generators are poor problem solvers.

The effective problem solver spends 50 percent of the time defining and analyzing the issue. Understanding the problem requires patience and determination. Such hard mental work deserves recognition – give yourself (and your group) a pat on the back for a clear definition of the problem.

If unsuccessful dieters spent as much time researching and understanding their weight problem as they spend trying out new solutions, they might discover the 'real' problem – and perhaps a solution that would work for them? (Quinlivan-Hall & Renner, 1994)

As the step develops, ideas become firmer. Some ideas may be clear at the start, the remaining becoming clearer.

There is a certain amount of creativity involved in the step, and so an exact prescription of what goes on in the step is difficult to give.

In problem areas where experience is available, then this experience can also aid the problem definition.

Problem statement

The words used in stating the problem have to be chosen carefully because they influence the direction later taken in attempting to find a solution. Connotations and people's natural tendency to artificially constrain their own thinking (unable to 'think outside the square') can contribute to the 'wrong' problem being solved, or the wrong path taken towards a solution.

Hall (1962) quotes the example of Arnold:

'If a man builds a better mousetrap, the world will beat a path to his door.' Is the intent that a path be beaten to your door? to build a better mousetrap? to get rid of mice? If the intent is to get rid of mice, is it to kill, exterminate, trap, electrocute, drown, scare-to-death, encourage emigration, encourage suicide, ... ? If extermination, is it by poisoning, gas, starvation, ... ?

On agreeing on definitions of key terms:

Peter sensed that members had diverse interpretations of the task. He suggested they take a moment to discuss and agree on the key terms of their problem definition. It read:

How to *develop* a *fund-raising* project for the *summer*.

Peter highlighted the key words and then facilitated the discussion of their meaning. Here are some points the group raised.

---

fund raising project

What will it be? Are we talking about a cookbook, a nature guide, a raffle, a series of Sunday events, a concert? What amounts do we hope/need to raise? Who will be involved in raising the funds? What group are we targeting with this fund-raising drive?

summer

When exactly does this have to be in place? By the start of the tourist season? How much time do we have between then and now?

develop

Are we here to come up with ideas or specific plans? Who will do the work?

In the end, they agreed on a much clearer definition. To decide whether to publish a cookbook that

    a. Appeals to a national audience
    b. Can be ready by June 1
    c. Can be done with no more than $2,000 start-up money
    d. Could raise at least $5,000 over two years

                                          (Quinlivan-Hall & Renner, 1994)

On defining a problem:

Two hikers are being chased by a grizzly bear. While on the run, one of the hikers reaches into his backpack and pulls out a pair of jogging shoes. The other hiker glances over and says, 'Why bother? This bear can outrun you even with those on.' His partner responds with, 'I don't need to outrun the bear, I just need to outrun you.'

                                          (Quinlivan-Hall & Renner, 1994)

Occasionally there may be a problem within a problem. For example, before a car wheel can be balanced, any damage to the wheel has to be fixed.

Needs analysis

Needs analysis or needs research establishes what the owner, consumer or stakeholder wants. This may involve a new product or service, changed performance level, changed costing, alternative functions etc.

Environment analysis

Everything not included in the system is the environment. Knowledge of the environment is necessary in terms of how it interacts with the system.

It could be argued that it is not possible to know everything, a priori, relevant to a problem. In such circumstances the problem solving proceeds with incomplete information. Research, education and experience can improve this situation.

One approach to environment analysis lists all possible inputs and outputs to the system.

Boundary conditions

Following from an environment analysis, and the establishment of a system-environment boundary, there will be boundary and initial and final (terminal) conditions on the system behaviour.

Constraints

Constraints or restraints restrict the number of possible solutions and fix many of the system properties.

Solutions that satisfy the constraints are termed *feasible, admissible* or *acceptable* solutions.

Example restrictions may relate to cost, weight, volume, appearance, operational/maintenance considerations, function, performance, acceptable risk levels, technology, time, capacity and tolerances.

Further examples include – the solution must:
- Be consistent with established policies and practices.
- Be acceptable to others; not pose problems to customers.
- Not increase the budget; within budget.
- Fit within existing workload.
- Achieve desired results by year-end.
- Not harm the (natural) environment.
- Use existing space, existing people and no more.
- Include certain people.
- Easy to implement.
- Have long-term impact.

For workplace problems, constraints might be classified as:
- Internal – that is, deriving from departments, groups, management, unions, ...
- External – that is, deriving from customers, government, ...

Assumptions

Some assumptions may be needed, but generally assumptions are to be avoided as much as possible.

4.4 SELECTING OBJECTIVES

Objectives may be part of the problem description, but it is more convenient to treat them separately because of their pivotal role in the whole problem solving process.

The objectives follow from a developed value system. What is wanted, desired or needed, that is valued, comprise the value system. These can be difficult matters.

Where alternative decisions or choices are possible, the objectives establish which is the preferred decision. The preferred decision is that which gives the 'best' values of the objectives; an extremisation (minimisation, maximisation) of the objectives is implied.

The difficulty arises because only very rarely do we find it necessary to make an explicit choice of ultimate goals. Our goals and intentions are formed during childhood, adolescence, and even maturity by instinctive likes and dislikes, by our parents, and by our experience in society. And when an explicit choice of goals is called for, there is no one kind of argument which must be presented to validate our stand. This is distinctly different from the situation in the logical sciences (logic and mathematics) where a proof is necessary and sufficient to validate an inference, or in the factual sciences (physics, biology, etc.) where observational data and predictive success suffice to substantiate factual claims.

By the statement that there is no one kind of argument to support a given choice of objectives, we mean that there is no body of theory to guide us in choosing objectives. This negative value is probably the most important fact about decision theory in general and value theory in particular. A consequence of this fact often can be observed in engineering reports, where the casualness with which objectives are reported contrasts sharply with the elaborate mathematical models, calculations and empirical data used to justify the choice of one system over another.

Yet it is much more important to choose the 'right' objectives than the 'right' system. To choose the wrong objective is to solve the wrong problem; to choose the wrong system is merely to choose an unoptimized system. (Hall, 1962)

Objectives may not be chosen in isolation, but rather with foreknowledge of an intended system/solution.

Example

Different value systems lead to the following example objectives:
- Profit (immediate, short-term, long-term).
- Production (product or service).
- Cost (income, economic feasibility, first cost, annual cost – includes the cost of money, depreciation and taxes).
- Quality (quantitative measures – zero defects, conformance to specification; qualitative measures – human response, psychological issues).

- Performance (overlaps with production and quality; figures of merit, efficiency factors, reliability, stability, response, speed, capacity, errors, ...; relate to particular systems).
- Competitive issues (retaining or capturing a market segment, affecting or damaging a competitor, minimising a competitor's profit).
- Compatibility (with existing systems, phase-in periods).
- Adaptability/flexibility (to a changing environment, ease of conversion, multiple uses).
- Permanence (obsolescence in all its forms).
- Simplicity/elegance (a subjective measure).
- Safety (including probability of system failure, value of consequent loss including loss of life and limb – itself debatable how it's measured).
- Time (including schedule, target dates, lead times, ...) – often implicit in other objectives.

In addition there may be combinations of these objectives, for example production per unit cost.

Some issues may be difficult to measure. Typically this relates to subjective issues.

Objectives may be peculiar to a particular individual or organisation because they derive from a value system. Individuals may not agree on what the objectives should be for any particular problem.

The objectives are stated early in the problem solving process, but may be modified, or refined or made more detailed as more information about the solution comes to hand. The selection of the objectives may occur in parallel with the solution.

Comment

> *There is no unique path to a good set of objectives. Even if one is given a set of objectives, there is no foolproof way to tell whether the set is good. The lack of a comprehensive approach to setting objectives is no excuse for not facing up to the problems of setting them. Neither does this lack justify the arbitrariness, imposition, dogmatism and absence of logical thought so frequently found in work on objectives.*
>
> *There is, happily, some useful and widely accepted theory and philosophy to guide work on objectives. Rather than guide us to the best objectives, the nature of most available theory is that it helps us to spot objectives that are wrong in some sense, or at least which are worse than some others that might be selected to guide the same action. Thus existing aids are rather weak, but it is safe to say that much better results can be achieved by using them than by approaching the problems of setting objectives blindly.*
>
> *Listed below are just a few suggestions that follow from the present state of knowledge about value system design.*
>
> *(a)   Put the objectives on paper. Get agreement that the words used are neutral and free of bias.*

(b)   *Identify means and ends. If this results in several chains of means and ends, 'position' the chains in order to locate objectives on the same hierarchical level, and to identify the different dimensions of the value system.*

(c)   *Test to see that the objectives at one level are consistent with higher level objectives. This is necessary to decide the relative importance of various subsets of objectives.*

(d)   *Test that the subset of objectives at each level is logically consistent. Inconsistent objectives signal the existence of trade-off relations. All of these relations will eventually need careful specification to allow for compromises.*

(e)   *Define the terms of trade for related variables. Sometimes all that is necessary is to find the derivative of one variable with respect to another, and to state the limits of the variables within which the trade is valid.*

(f)   *Make the set of objectives complete. The use of experience in similar problems is one way to satisfy this criterion. However, this operation generally continues to the end of design because it is impossible to foresee all consequences of the physical system and cover them with objectives.*

(g)   *Give each objective the highest possible level of measurement. Recognise that some objectives are not measurable on the highest level of measurement scales. Usually the members of certain subsets of objectives can at least be ranked by importance.*

(h)   *Check the objectives to see if each is physically, economically, and socially feasible. State the limiting factors.*

(i)   *Allow for risks and uncertainties by various available techniques and by selecting an appropriate decision criterion.*

(j)   *As a step in settling value conflicts, isolate logical and factual questions from purely value questions. This frequently calls for the use of experts.*

(k)   *Settle value conflicts. Have all interests represented. Use tentativeness. Avoid dogmatism, dictatorial methods and premature voting.* (Hall 1962)

4.5 GENERATING ALTERNATIVES/SYNTHESISING SYSTEMS

4.5.1 General

Although possible solutions may have been in the problem solver's mind from the outset, strictly the problem definition step and selecting objectives step are setting up the problem, that is, *what* is the problem. The step of synthesising systems or generating alternatives now looks at possible solutions. This is the *idea generation* step.

Synthesis involves putting the parts together to form a whole. Selection of the parts and how they interrelate is implied.

This synthesis process may be routine for established systems/problems. The parts and their interrelationships are predefined by custom or accepted theories and practice.

Other systems/problems may too follow a logical approach.

The other extreme for the synthesis process is where it is extremely creative. *Inventions* follow this path. Something new, but generally along the lines of something else already existing, involves much less creativity input because the synthesis process is reversed in order to understand the result, and then reapplied with a small variation.

Most synthesis is carried out following previously developed approaches to obtain results similar to that already existing. Previously developed approaches are extrapolated or interpolated.

Once a mathematical model exists in a discipline, synthesis for systems/problems follows along the same lines, tailored with minor modifications to particular systems/ problems. The minor modifications might relate to changing boundary conditions or parameter values, or assuming some parts of the model are less dominant than others and hence can be discounted.

The end result of systems synthesis is to come up with a range of possible solutions that satisfy the problem definition.

4.5.2 Idea generation

There are a number of methods that can be used to generate ideas including:
- Taking an idea census.
- Functional design.
- By delineating subsystems.
- Pure creativity.

Never reject an alternative because at first it seems foolish or far fetched. A useful technique at this stage is to forget one's critical abilities in the search for ideas. This applies whether one is seeking ideas from others or trying to get ideas from himself. Imagination and evaluation are antithetical to some extent, and these activities can be carried on separately at this stage. Criticism applied too early may inhibit the free flow of ideas. The systems engineer in seeking alternatives from others might well cast himself in the role of the gullible listener – not criticizing or evaluating alternatives, but encouraging free exposition and promotion of all ideas, not only current ideas but also those that have lain idle in some long-neglected file. Criticism applied at the proper time, and in the proper spirit, may stimulate another flow of ideas.

Sometimes ... the systems engineer must rely on his own imagination and invent a system plan, call upon experts in the relevant field, or abandon the project. It often happens, of course, that the systems engineer does invent, if only because he may be the first technically trained person to encounter the problem. In this case the systems engineer, whose main function is reaching unbiased conclusions, has a new hazard; he must not allow himself to be committed emotionally to his own alternative if other alternatives should appear. He must continue the search for other alternatives to place in competition with his own; he must strive for more than one way of doing a job, if only to dilute his committal to one way.

The systems engineering method may contain within itself checks against such dangers. For one thing, the systems engineer may at the conclusion of planning write a report of his studies and thereby expose his plan to wide criticism. For another, once a report is written, the entire project may be transferred to another individual or group in a different department to pursue the next phase. This procedure has among its advantages the strengthening of the broad conception and eliminating from it the quite natural bias given to it by its inventor. Sometimes this advantage is paid for in loss of time while a new group goes over some of the same ground. (Hall, 1962)

Taking an idea census

All known possibilities or alternatives are collected from any available source. With a broad problem statement, many alternatives could be expected to be found; a narrow problem statement could be expected to severely restrict the range of alternatives.

There is commonly *the do-nothing alternative* and *the status quo alternative*.

In some situations it may be that there is no existing solution, or no analogy can be found.

Functional design

> *Of all the techniques for aiding the creative approach in planning and design, none is more significant than functional synthesis and functional analysis. Sometimes this technique is oversimplified and called 'block diagram design.' The technique starts with a statement of boundary conditions, and desired inputs and outputs, and proceeds to a detailed list of functions or operations which must be performed. Then these functions are related, or synthesized, into a system model showing essential logical and time relationships. At this stage only the requirement of reliability applies; all that is wanted is a combination that works. Later, the model is analyzed relative to performance and other objectives for purposes of optimization. The functions usually can be permuted into a number of alternative arrangements for accomplishing the [requirements]. (Hall, 1962)*

> *In the early stages of planning, discussion of hardware sometimes confuses the distinction between how to do the job with what the job is. To demonstrate a feasible plan it is usually necessary to consider choices among devices; when these choices are made, additional functions must be added to the list. Put another way, some system functions – the minimal or invariant ones – arise because of the nature of the job to be done; other functions arise because of the techniques or devices chosen to do the job. The precaution against excessive preoccupation with hardware cannot be taken too far because the distinction between minimal and nonminimal functions is not sharp; the extent to which functions are necessary and which functions can be combined depends strongly on the state of the art.*

> *Functional planning applies with greatest effect to the area of design for increased or new function as distinguished from design for lower cost, higher performance, or greater saleability. (Hall, 1962)*

Delineating subsystems

The function to be performed by the system determines many of the subsystems and subsystem boundaries. Subsystems themselves may be further subdivided in a form of *progressive factorisation*.

Often the subsystems so determined are still too large, so they are subdivided repe-
atedly according to function or other principles noted below.

Problems of interconnecting subsystems generally are reduced when the number
and/or number of kinds of inputs and outputs for a subsystem are minimized. This is
as true in the design stages, where different design groups have to make the various
subsystems compatible, as it is during factory assembly and installation.

Another principle of sectionalization is to minimize the number of interactions
between subsystems; this does not always give the same results as simply minimi-
zing the number of inputs and outputs. For example, electronic subsystems tend to
interact unfavourably through common power supplies. An extreme design coun-
termeasure which is taken in some missile systems is to provide separate power
supplies for each subsystem, thus tending to minimize both the number of inputs and
interactions.

Delineating functions or subsystems by geographic location is an important and
frequently used criterion. ... The choices of functions and components that belong
at one location are not always so obvious. ...

Sometimes sectionalization is influenced by the ability of a given specialist or
specialist group to do the design work, and sometimes by the need to keep a design
group small enough to manage without excessive supervision or reporting. Such
criteria are not usually as important as the preceding criteria, unless the utmost in
speed and efficiency is paramount. (Hall, 1962)

Pure creativity

... criticism inhibits ideation. Sociologists and psychologists have shown that many
additional factors inhibit the free flow of ideas. When there is no model of a previ-
ously designed system to adapt, no semilogical synthesis techniques, no authority to
turn to, the last resort is one's own imagination. In this case, it sometimes helps to
know what the inhibiting factors are and to try out a few techniques for overcoming
them. (Hall, 1962)

4.6 ANALYSIS/ANALYSING SYSTEMS

Each alternative generated can now be analysed relative to the previously established
objectives and constraints, leading to the selection of the optimum solution in the next
step.

The analysis step may also provide information that feeds back to the synthesis and
objectives selection steps. It may uncover previously unthought of issues. The optimisation process may contain many such feedback loops.

The form that the analysis takes reflects the form of the objectives and constraints.
For example if the objectives and constraints are in money terms, some cost/financial
analysis is undertaken.

Any analysis tool may be appropriate. Such tools may range from being very mathematical through to being qualitative. An example of a qualitative analysis is where people
might be asked to attach attributes to each of the alternatives.

Analysis might be, for example, with respect to:
- Practicality.
- Realism.
- Cost-effectiveness.
- Ease of implementation.
- Consistency with something else, perhaps existing.

Uncertainties

Uncertainties might be handled through probabilities, obtained either objectively (for example, historical data) or subjectively (for example, an expert's opinion or belief).

Sensitivity analyses also assist in this regard.

4.7 SELECTING THE BEST ALTERNATIVE

The best alternative is the optimum solution.

Where there is only a single objective, the selection is straightforward. For example, if minimum cost is the objective, all alternatives are evaluated for cost and the one that gives least cost is chosen.

This example is also deterministic.

The situation becomes more difficult when multiple objectives and/or uncertainty are introduced. With multiple objectives, some subjectivity, reflecting the decision maker's values, has to be introduced. With uncertainty, the calculations become more involved and are restrictive for any problem other than small ones.

This step is an *evaluation* phase. Some writers intertwine analysis and evaluation, and this confuses the methodology. Evaluation is with respect to the objectives and constraints.

Matrices might be used as an evaluation tool. The objectives/constraints and the alternatives represent the two axes.

	Objectives					Constraints		
	1	2	3	4	5	(i)	(ii)	(iii)
Alternative A								
Alternative B								
Alternative C								
Alternative D								
...								

Evaluation matrix

Sorting of alternatives may be done in a number of ways:
- *Duplication and connections.* Alternatives that duplicate others might be deleted. Alternatives with similarities might be combined.
- *Rank ordering.* Alternatives are ranked and the lowest ranked ones eliminated.
- *Categories.* Alternatives are placed into different categories. The categories may focus the selection process.
- *Candidates for deletion.* Some alternatives will not warrant further consideration and can be deleted.

Voting may be carried out in different ways. This applies to group problem solving. Each alternative might be ranked by each group member and collated across all people. The alternative which achieves the best composite ranking is then selected.

EXERCISES

1. What might be used as indicators of a problem existing?

2. How do you know that you have identified the correct problem, as opposed to a near or related problem? In particular, do hidden agendas affect the problem definition activity?

3. What is your view on spending 50% of the problem solving time on problem definition?

4. How do you get the correct balance between too broad a problem definition – and hence many solutions – versus too narrow a problem definition – and hence too few solutions?

5. Many people confuse objectives and constraints. Why is this?

6. Would you expect objectives to be more readily stated for an organisation or an individual?

7. Do individuals know their own value systems well enough to explicitly state their objectives?

8. Suggest some ways (other than those given above) alternatives could be generated.

9. How much creativity is exhibited by people, or is it the case that most of the time people copy off someone else's ideas or experiences?

10. Why is it important to keep the analysis step separate from the next (evaluation) step? Or doesn't it matter?

11. Of all the alternatives generated, most will be rejected in any particular problem solving situation. Is there any value in taking note of these rejected alternatives for future problem solving activities, or is it better to start the creative thought processes from anew each time?

Should any review of the problem solving process account for the rejected alternatives?

12. What approaches (other than those mentioned above) to selecting the best alternative can you suggest?

APPENDIX 4/1

Value Management

4/1.1 GROUP PROBLEM SOLVING

Groups are often established for the purpose of solving problems. It might be called a meeting or some more up-market group name such as focus groups, quality circles, management teams, task forces, boards, committees, employee groups, ... To advance the problem solving process a person, possibly called a facilitator or chairperson, is appointed. The facilitator organises the problem solving process, provides structure to something that can become chaotic, and where necessary focuses the group thinking.

Connected issues relate to the physical environment, the attitude of individuals and the group, problem solving culture, and agenda and personal time management.

The problem solving steps are similar for individuals and groups. Groups, however, introduce additional matters relating to how the individuals interact and respond to each other and each other's ideas. Personality clashes and procedural wranglings can interfere with the group process. Meetings, where members become frustrated because of poor chairing, unclear goals or lack of commitment by others, have been experienced at some time or other by most people; the meetings are plainly nonconstructive.

Value management

Value management/engineering/analysis attempts to find solutions which are cheaper but still perform the required function. Value management workshops involve input from various specialisations that impinge on the problem; they are facilitated and run as problem solving workshops.

A more liberal, but minority, interpretation by some people of value management, is that it is a process of reconciling the interests of disparate groups. As such it is also a problem solving exercise. Commonly community consultation workshops, surrounding an impending project, might be called value management workshops in this sense.

4/1.2 VALUE MANAGEMENT OUTLINE

4/1.2.1 Introduction

Value management is popular in project management for a diverse range of situations. Originally conceived as a way of economising on resources or finding alternatives, it finds application at all phases of a project, from concept through to termination. Central to the approach is a systematic analysis of function and an examination of alternatives. As such it can be shown to be but a special case of systems engineering/problem solving methodology. Accordingly, the need for the introduction of the terminology 'value management' and the industry sector that promotes itself as value managers is questioned, when established tools and approaches exist, albeit with a less glamorous title. The misuse of the terms 'management' and 'value' in its title are also questioned – it is suggested that the title is a misnomer.

> *Value management is a structured, systematic and analytical process which seeks to achieve all the necessary functions at the lowest total cost consistent with required levels of quality and performance.* (New South Wales Govt, 1992)

Its origins trace back to about the 1950s when it was known as *value analysis* or *value engineering*. Then it was a design review or 'second look' approach to proposed or existing designs. Its area of application has enlarged over the intervening years such that it now encompasses not only design reviews but also, for example, feasibility studies, an examination of project objectives, and conflict situations.

Constructability or *buildability* studies are but special cases of value management applied to construction or building projects. Constructability analysis might be defined as:

> *A review of plans and specifications from the viewpoint of the constructor, which would include opportunities for prefabrication, preassembly, modularisation, special construction methods and other considerations aimed at lessening the cost or improving the completion schedule.* (Barrie & Paulson, 1992)

Significant cost savings have been reported through the use of value management by a number of writers, for example New South Wales Govt (1992).

Central to value management is the analysis of function from a whole system viewpoint and the proposing or generating of alternatives. It identifies wastage, duplication and unnecessary expenditure, and can assist in testing assumptions and needs. The whole system or holistic approach avoids conventional compartmentalised thinking and obtaining locally optimal or suboptimal solutions, at the expense of the desirable globally optimum solution.

This *raison d'etre* is no more than a paraphrase of that in a systems engineering approach which incorporates problem solving methodology.

From a project viewpoint the application of value management achieves best results if applied in the early concept and development phases, but it has applicability at all phases and in all tasks.

4/1.2.2 Value management process

Different writers suggest slightly different steps, and terminology for the steps, in the value management process or 'value study'. The approach suggested in New South Wales Govt (1992) is a reasonable consensus and is repeated here:

Information
Identification and testing of program or project rationale from the perspective of stakeholders' positions.

Function analysis
Identification and ranking of primary and secondary functions and their associated cost and worth relationship.

Ideas generation
Generation of value improvement options through innovation and alternative means of achieving the required function.

Evaluation
Sorting and prioritising value improvement options to identify viable alternatives.

Action plan
Identification of actions required to achieve value study outcomes and provide an ongoing management framework for project progression.

This is followed by reporting.

The duration of a value study may be anything from a half day to several weeks depending on the matter being studied. Where prior preparation is necessary, the time span is at the top end of this scale.

An integral part of a value study is a workshop or gathering of interested individuals and stakeholders. The workshop acts as a group problem solving session with its associated benefits achieved through combining knowledge and accessing the group dynamics. Participants for the workshop are selected such that they represent all disciplines impinging on the problem. Suggested numbers of participants vary from writer to writer, ranging from about 5 up to about 15, but at the top end the management of the workshop could be expected to be more difficult and the effectiveness of coming up with new ideas could be expected to decrease.

4/1.2.3 Comparison

The definition of system is seen to be sufficiently broad to cover any entity considered in a value management study. For example in a study of a facility, the facility is the system; and in a study of a procedure, the procedure is the system.

The process, steps and information sought in a value management study are seen to translate directly from the same in the systems engineering/problem solving approach (Figure 4/1.1).

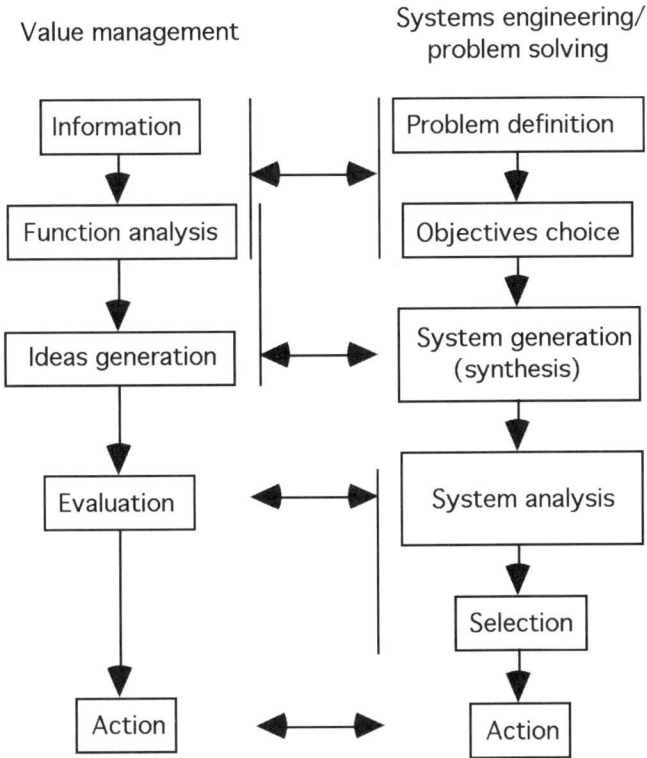

Figure 4/1.1 Value management and systems engineering/problem solving equivalence.

The question, then to be asked, is what differences between a systems engineering/ problem solving approach and a value management approach can justify the introduction of a new, and stronger critics would say, pseudo-new, field. Definitely it is not justifiable just because some practitioners in value management have had educations that have not demanded rigour or the rigour of systems engineering/problem solving; some more recent industry qualifications are wanting in this respect.

Contributions

Some minor contributions appear to have been made by value management writers.

One contribution involves the distinction between primary and secondary functions of a system component or the development of a function hierarchy, for example Table 4/1.1. This appears to be a useful idea. Systems engineering looks at purpose and function of components but no ranking is usually carried out, although the approach does not exclude the possibility of examining a function hierarchy. Related to this contribution is the development of FAST diagrams, FAST here standing for function analysis system technique.

Figure 4/1.2 shows a typical FAST diagram.

Table 4/1.1 Example function analysis.

System/ component	Function	Kind
Pen	Form shapes	Basic
Casing	Holds ink/nib	Secondary
Nib	Forms shapes	Basic
Ink	Colours shapes	Basic
Cap	Protects nib	Secondary
Clip	Attaches to pockets	Secondary
Markings	Identifies pen	Secondary
	Improves appearance	Secondary

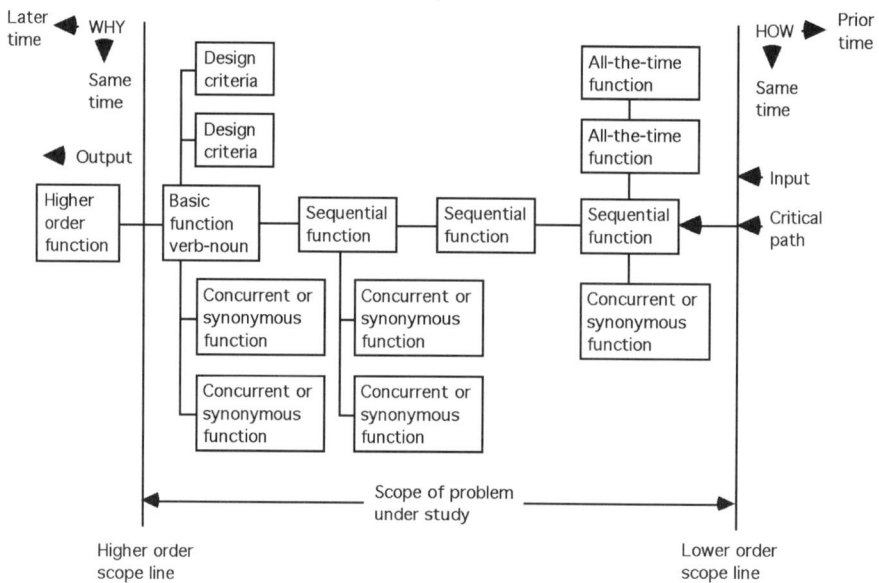

Typical FAST diagram

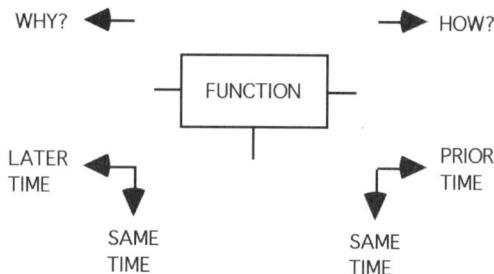

Figure 4/1.2 Diagrammatic description of FAST (after Dell'Isola, 1982).

The functional examination from a system perspective is fundamental and involves an analysis of the combination of interrelated, interdependent, or interacting elements forming a collective entity and serving a common set of [goals].

If the subject project is a building, it of itself is a system comprising many parts serving a common function. Within the building there are many subsystems such as the air-conditioning, transportation etc.

Taking a wider view of a system, if the subject building is a hospital it forms part of the health system; a bus depot forms part of the transportation system; and a dam forms part of the water supply system. (New South Wales Govt, 1992)

4/1.2.4 Closure

Value management can be shown to be subsumed within the classical problem solving approach of systems engineering. Accordingly it is difficult to see what benefits introducing a new term, namely value management, has. As a secondary issue, the choice of the words 'value' and 'management' leaves much to be desired as neither or both describe with accuracy what is practised, and they give a misleading view to purchasers of the service.

The worth of value management type studies is not questioned, only the need by some to dress up existing methodology and practices in a fashionable way.

EXERCISES

1. Why do people need to introduce glorified titles, such as value management? Why not treat it as no more than problem solving? Are people deficient in their education, and hence don't realise they are reinventing the wheel? Or is there a cynical commercial motive at the base of every new fancy management title or fad?

2. What is 'value'? Why is the term 'management' used in the technique called 'value management'?

3. In group problem solving, what would be a recommended number of people? On what factors does the choice of number depend?

4. A FAST diagram proceeds from left to right based on the question HOW? and from right to left based on the question WHY? Does this imply some relationship between HOW and WHY?

APPENDIX 4/2

Risk Management

4/2.1 PREFERRED DEFINITIONS

Risk is a pervasive part of all actions. It would seem on the surface that the term 'risk' is a simple well-understood notion. However, its definition is elusive, and its measurement is controversial.

In the literature, the word 'risk' is used in many different meanings with many different words such as hazard or uncertainty. It is found that there is no uniform or consistent usage of the word risk in the literature. In addition, most definitions of risk have focused only on the downside associated with risks such as losses or damages, and neglected the up side or opportunity such as profit or gains. This work recognises both sides of risk i.e. the downside and its counter-opportunity. (Al-Bahar & Crandall, 1990)

For this book, the definitions are adapted from Al-Bahar and Crandall (1990):
Risk The exposure to the chance of occurrences of events adversely or favourably affecting the project/business/... as a consequence of uncertainty.

Risk event/source/factor What might happen to the detriment or in favour of the project/business/...

Uncertainty of an event How likely the event is to occur; that is, the chance of the event occurring. A sure or certain event does not create gain or loss.

Potential loss/gain Loss is used as a general term to include personal injury and physical damage, and gain to include profit and benefit. There is some amount of loss or gain involved as a consequence of an event happening.

Risk management A formal orderly process for systematically identifying risk events, analysing and evaluating potential consequences, and responding to obtain the optimum or acceptable degree of risk elimination or control.

Symbolically, we could write this as:

Risk = f(Uncertainty of event, Potential loss/gain from event).

From this definition, uncertainty and potential loss or gain are necessary conditions for riskiness. It may seem strange to refer to uncertainties about potential gains as risks. However, even in situations of potential gains, uncertainty is unattractive since the knowledge of the exact gains is unknown, and contractors are reluctant to give credit to an unknown gain. (Al-Bahar & Crandall, 1990)

4/2.2 ALTERNATIVES

The above is the preferred definition of this writer for the term risk. However, as noted, it is not unanimously accepted. On reading the risk management literature and talking with practitioners, it has to be kept in mind that people, unfortunately, cannot agree on a single definition of risk. Many people are loose in their usage of the term risk.

Some examples
1. Some define risk as a probability, for example:

The chance of something happening that will have an impact upon objectives.

To this writer, such a definition is unacceptable. Firstly, consider an extreme event such as a cyclone. It has a chance or probability of occurrence, and it impacts on whatever activities you are engaged in. Risk management therefore must be the management of the chance or probability of occurrence of cyclones. Apart from cloud seeding and similar meteorological methods, there is no way that the chance of cyclones occurring can be managed. The chance of occurrence of cyclones is up to mother nature. Perhaps users of such definitions are referring, not to the initiating event, but rather some consequent event when it refers to 'something'. But the definition is unclear.

2. Some define risk as a possibility, for example:
Risk is the possibility that an expected outcome is not achieved or is replaced by another, or that an unforeseen event occurs.

Possible outcomes are always present where uncertainty exists. The management of possibilities is not believed what is intended, because of the lack of ability available to do this. Or is the intention to manage the size of the sample space?

3. Dictionaries give lay meanings of words and generally are not useful when it comes to working technical definitions. For example:
1. the possibility of incurring misfortune or loss; hazard

2. Insurance
   a. chance of a loss or other event on which a claim may be filed
   b. the type of such an event, such as fire or theft
   c. the amount of the claim should such an event occur
   d. a person or thing considered with respect to the characteristics that may cause an insured event to occur

Only a definition approaching (2c) would appear to be useable within a consistent approach to risk management.

But be aware, all definitions and others are used. Sometimes several definitions are loosely used by the one person in the one discussion. Flip-flopping from one definition to another seems a characteristic of a number of risk management practitioners.

4/2.3 OTHER ISSUES

Loss and Gain

Some people only think of risk in terms of its downside, that is a loss, damage, extra cost, extra time etc.

However there is another view that thinks of risk in terms of both the downside and the upside. Just as things can possibly go wrong, so events can occur that lead to some gain.

Generally, it has been observed, that most people undertake risk studies in an attempt to deal with potential losses. However the potential gains should also be part of the studies.

Risk or risk event

It appears some of the confusion over the definition of risk may arise because of some people using the terms *risk event* (or equivalent) and *risk* interchangeably.

This is particularly so in what is termed 'risk identification'. In fact what is meant by 'risk identification' is usually 'risk event identification'. Figure 4/2.1 tries to differentiate between a risk event and a risk. The process of risk management generally follows Figure 4/2.1.

Identification starts in the left hand box, or in some cases may start with some idea in the right hand box and then work back to the left to establish what is in the left hand box.

4/2.4 THE RISK MANAGEMENT PROCESS

See Figure 4/2.1.

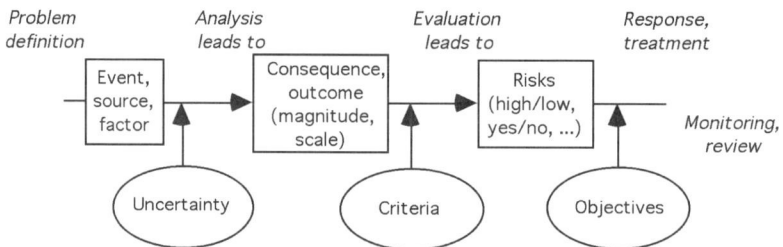

Figure 4/2.1 The risk management process.

What is termed 'risk analysis' might be more appropriately called 'analysis leading to potential consequences/outcomes'. 'Response' or 'treatment' deals with influencing the transformations or the initiating events, in order to address the risks detected.

Successful completion of a project, or the successful undertaking of a business or other endeavour, could be expected to depend on identifying all the risk events or sources associated with the project, business, endeavour, ..., analysing the risk situation, evaluating the consequences and responding to the associated risks.

Projects, businesses, ... generate their own characteristic way of looking at risk which is different to other situations involving risk. Each project, business, operation, ... is unique and generates its own set of problems. As well, historical data may not be available to assist the analysis process; actuarial type calculations may not be possible.

Risk exists because of uncertainties. Uncertainties may arise from quite diverse causes including the regulatory process, natural hazards, monetary matters, unproven technology, management matters, resource availability, industrial relations problems and so on.

Failure to deal effectively with the risk events or sources will impinge on the cost, time and quality outcomes of a project, business, ..., on safety issues etc.

4/2.5 A SYSTEMATIC FRAMEWORK

Systematic problem solving or a systems engineering approach may be broken down into a number of operational steps:
• Defining the problem (including issues and situation).
• Selecting objectives.
• Generating ideas, alternatives.
• Analysing ideas, alternatives.
• Selecting the best alternative.
• Action.

and where the problem situation is ongoing, there is an additional step of:
• Monitoring and reviewing.

The risk management process is essentially the same, though different terminology might be used and the number of step classifications might be different. Some commonly used terminology is:
• Establishing the context.
• Identifying (risk events).
• Analysing the risk situation.
• Evaluation of consequences/assessment.
• Responding/treatment.
• Action.
• Monitoring and reviewing.

Figure 4/2.2 shows the relationship between the problem solving process and the risk management process.

As in the problem solving process, there may be feedback between steps in attempting to refine the process or as secondary influences materialise.

The first step establishes the context for the rest of the process. Part of this is establishing the criteria by which risks will be assessed and the objectives influencing the response.

Identifying risk events requires the same type of thinking as generating ideas/ alternatives of the problem solving model. A certain amount of creative thinking is required in this step for all but the most straightforward situations.

The analysis step may be carried out qualitatively or quantitatively using whatever technique and approach is appropriate.

The evaluation is carried out with respect to the previously stated criteria. This might involve some ranking and prioritising of consequences, and even the dismissal of consequences regarded of little effect according to the criteria.

Establishing the context

In this definitional step, the following are established:
- The *context* of the risk management problem.
- The *criteria* by which risks are to evaluated. (For example, high risk might correspond to a large exposure in dollars; consequences that impact humans may be regarded as primary risks, while consequences related to budget and operations may be regarded as secondary risks.)

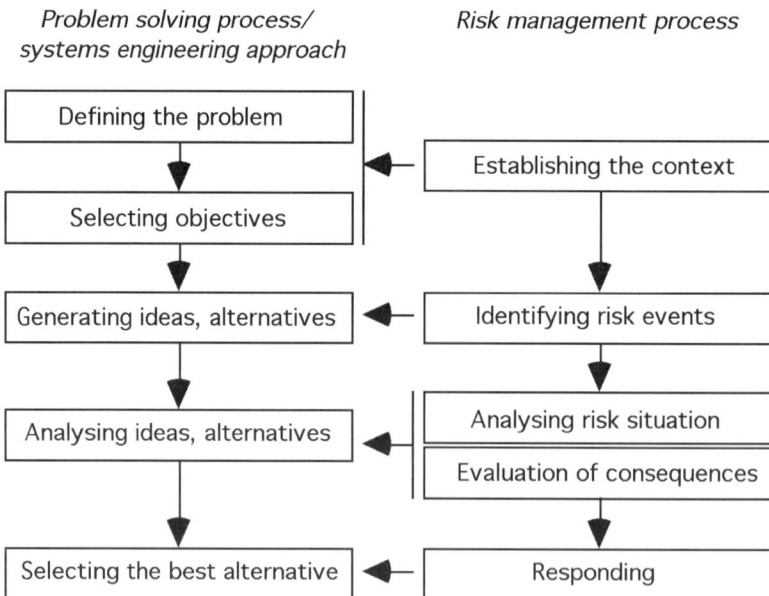

Figure 4/2.2 Problem solving process and risk management process compared.

- The *objectives* by which the managerial decisions (response/ treatment) are to be made. (For example, minimum initial and ongoing cost.) (Where these relate to final (terminal, end) states, they may be referred to as goals or targets.)
- *Constraints* on the decision making. (For example, certain consequences may be totally unacceptable.)

The middle two items, and possibly the fourth item, come from value judgements on the part of the risk manager, and hence introduce subjectivity into the whole process.

Identifying risk events

Many writers call this step 'risk identification'. In fact, what is generally meant is 'risk event identification' or identification of the sources of risk. Certainly potential consequences or outcomes are considered, but only as an exercise in then working backwards to establish the underlying risk event or source – the event that results in the consequence or outcome.

Many people and texts are loose in this matter.

In safety matters, this step might be referred to as *hazard identification*, or developing a *hazard scenario*.

The identification of risk sources, events, ... may be carried out with the help of checklists, interviews, brainstorming sessions or personal and corporate past experiences, and other more specialised approaches.

There is a Pareto effect at work here, in that 80% of the losses/gains are the result of 20% of the possible events.

Analysis (of risk situation)

Many writers call this step 'risk analysis'. In fact, what is meant, is the conversion of the risk event information into information on consequences or outcomes. A more appropriate term to 'risk analysis' might be 'potential consequence/outcome analysis'.

Many people and texts are loose in this matter.

Commonly, analysis may be carried out in two parts. The first part is a *qualitative* analysis that carries through a subjective view of consequences. The second part is a *quantitative* analysis that presents a more definite view. For cursory risk management, only the qualitative analysis may be carried out.

Often identification is intertwined with analysis and evaluation/ assessment and the three proceed hand-in-hand.

A quantitative analysis would generally employ some numerical approach, often using a computer. It revolves around establishing quantitative estimates for costs, times, ... and uncertainties.

A prior qualitative analysis is recommended in order to obtain an overall understanding, seeing where all the components fit together and their relative contributions to the big picture. Such an analysis may be carried out such that low impact consequences are excluded from a more detailed study. If no more is done, at least a qualitative analysis is considered essential.

Evaluation of consequences

This step may also be referred to as *risk assessment.*

Responding

Having established the risks with their various degrees of importance, it is now necessary to develop a response. This step may also be called *risk treatment.*

This is the decision making step of the risk management process. Decisions are made based on the objectives and constraints delineated earlier in the process.

Low or minimal risks may be accepted without further consideration, but the situation might be continuously monitored to ensure it remains acceptable.

Part of the response is the development of a plan to deal with specific and overall risk issues.

Two types of response may be recognised (Figure 4/2.3):

Trade-offs may be considered in terms of responding at different points in time and the different consequences associated with responding at different times.

On risk source identification, analysis, and risk response:

> *'I was wondering what the mouse-trap was for,' said Alice, 'It isn't very likely there would be any mice on the horse's back.' 'Not very likely, perhaps,' said the Knight; 'but if they do come, I don't choose to have them running all about.' 'You see,' he went on after a pause, 'it's as well to be provided for everything. That's the reason the horse has all these anklets round his feet.' 'But what are they for?' Alice asked in a tone of great curiosity. 'To guard against the bites of sharks,' the Knight replied.*

Lewis Carroll, Through the Looking Glass

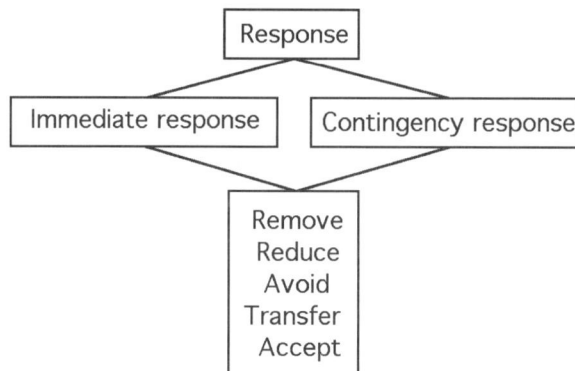

Figure 4/2.3 Response options.

The following tale highlights several aspects of risk management – the difficulty of fore-seeing all risks, and the lack of certainty involved in decision making where risks are involved.

The Lady or the Tiger?
(source unknown)

Which door to open? A young man could open either of two doors. If he opened one, there came out of it a hungry tiger, the fiercest and most cruel that could be procured, which would immediately tear him to pieces. But if he opened the other door, there came forth from it a lady; the most suitable to his years.

The first man refused to take the chance. He lived safe and died chaste.

The second man hired risk assessment consultants. He collected all the available data on lady and tiger populations. He brought in sophisticated technology to listen for grow-ling and detect the faintest whiff of perfume. He completed his check lists. He developed a utility function and assessed his risk averseness. Finally, he opened the optimal door. And was eaten by a low probability tiger.

The third man took a course in tiger taming. He opened a door at random and was eaten by the lady.

EXERCISES

1. Write your understanding of the term risk.

What is the dictionary definition of risk?

Is there a difference between a lay person's use of the term risk and the one adopted here?

2. There is much confusion in the literature and public discussion on a definition of risk. Interestingly it doesn't seem to bother practitioners and many academics. Why is this the case? Is a rigorous definition not important? How can people talk to each other meaning different things for different words?

3. Australians use the term 'no risk' in everyday language to imply assent. 'to take or run a risk' implies proceeding without regard to the possibility of the danger involved. Are these proper usages of the word 'risk'?

What do 'a calculated risk' and 'to take unnecessary risks' mean?

Risk is usually associated with danger and peril. Is this a proper usage of the word 'risk'?

What is the relationship between 'hazard' and 'risk'?

One writer defines safety as 'the freedom from the totality of risk'. Comment on this usage.

What is the relationship between risk and: a brave person; an entrepreneur; great explo-rers like Columbus and Cook; explorers of space?

Insurance policies use the terminology 'sudden and unforeseen' when describing events. Does this imply uncertainty or ignorance?

4. Safety factors have been and are used in engineering design. For example, in structural design, a factor of 1.5 for dead load and a factor of 1.8 for live load were commonly used. How do safety factors express the underlying uncertainty in loads and material characteristics? Why is a larger safety factor used for live load than dead load? Would you expect the magnitude of the safety factor to vary with the degree of uncertainty? How can you relate safety factors to risk management?

5. Projects are comprised of phases. Do later project phases inherit all the risks from previous phases, such that risks accumulate as the project goes along?

6. Consider a project or work matter with which you are currently involved. What exposures would you anticipate?

How would you estimate the probability of occurrence of the sources of the exposures that you have anticipated?

How would you deal with such exposures?

7. The risk management process represents a special case of the problem solving process. How do such special processes come about? Why do not people start with the existing and more generic problem solving process and either use this or specialise this? Or do people have trouble starting with the general and going to the specific; is it easier to approach any situation from the specific?

Why is it that, having established that the risk management process is a special case of the problem solving process, people don't like the niceties of the generality offered? Some people essentially ignore or pretend this fact doesn't exist. Why cannot they see that it adds to the understanding of the risk management process?

8. Certain people in life appear to be good risk managers. What are the characteristics of a successful risk manager?

9. Objectives or criteria result from value judgements. How are these established for organisations, as opposed to individuals?

10. Why is it that people can quite happily use the terms 'risk' and 'risk event' interchangeably, when they clearly mean different things? How rigorous are you in your usage of terminology?

11. Gross movements in currency values. As part of doing business with other countries, selection of the preferred currency of payment and sensitivity analysis to movements in the value of the currency are standard decisions and actions that are taken.

What is not usually considered is the impact of gross currency movements such as occurred in 1997 in some Asian countries. The values of currency changed by 35% and more.

Such movements in currency values appear not to have been able to be foreseen. That is, the risk event or source would not have been identified.

Where does this leave the practice of risk management, if the cause of a major risk is not identified?

12. In what circumstances would you anticipate that a qualitative analysis by itself would not be sufficient? At what point does a quantitative analysis become essential?

13. Think about previous projects or work matters with which you have been involved. List some of the things that went wrong or astray because the consequence was not foreseen in magnitude or type.

14. Evaluation or assessment is based on previously established criteria. How do you deal with the situation involving multiple criteria? Or are multiple criteria not possible in the risk management process? If they do exist, give some examples of possible multiple criteria.

15. Consider an everyday event in your life, for example, crossing a street busy with cars. The risk relates to your getting injured or even possibly killed.

What practices can (i) you personally adopt, and (ii) the road authorities/legislators adopt, with respect to pedestrians, to:
• Remove the risk.
• Reduce the risk.
• Avoid the risk.
• Transfer the risk.
• Accept the risk.

APPENDIX 4/3

Work Study

4/3.1 INTRODUCTION

Work study attempts to create in people a questioning state of mind in the way they view their current and future work practices. This aware state of mind applies to all phases of a project and on a continuing basis in an organisation. Work study involves the critical and systematic analysis of work with an ultimate view to improvement and the elimination of any nonproductive elements. Work study is made up of *method study* and *work measurement*. Method study breaks down work into its component elements and questions the purpose and need of each element; it involves, amongst other things, recording information on the work, critically examining the facts and the sequence, and developing alternatives. Work measurement is concerned with the time to perform work. Work study is an established technique that has wide applicability.

4/3.2 WORK STUDY OUTLINE

Work study may be broadly defined as an analysis of the use of people, materials and equipment in work tasks, and an associated attempt at improvement and elimination of waste. Besides the financial incentives, it attempts to create an attitude of mind about the effective use of people, equipment and materials.

The agreed definition of work study ... is that it is a generic term for those techniques, particularly **method study** *and* **work measurement***, which are used in the examination of human work in all its contexts, and which lead systematically to the investigation of all the factors which affect the efficiency and economy of the situation being reviewed, in order to effect improvement.*

The [goal] of work study is to assist management to obtain the optimum use of the human and material resources available to an organisation for the accomplishment of the work upon which it is engaged. Fundamentally, this [goal] has three aspects:
1. the most effective use of plant and equipment;
2. the most effective use of human effort; and
3. the evaluation of human work.

The definition of work study is that given in several sources; see for example International Labour Office (ILO, 1969). This quotation follows Currie (1959).

Work study could be said to be as old as work itself. Formalism however has been attempted as far back as the late 1700s by J.R. Perronet, with later contributions by R. Owen, C. Babbage, F.W. Taylor, H.L. Gantt and F. and W. Gilbreth among others. Of these, perhaps Taylor's 'time study' and his 'scientific management', Gantt's 'Gantt chart', and Gilbreths' 'motion study' are best known to most students of management. Work study appears to have reached its peak in popularity in the 1940s and 1950s with a steady interest ever since.

The intent in most work study endeavours is to improve productivity. Productivity is the ratio between output and input. It may be described in terms of output (production) per resources used to give that output. The resources may be labour, materials, equipment or services. Through increased productivity may come higher profit, lower prices, higher wages etc. The attack on improved productivity comes from several angles including an examination of the basic processes, plant, equipment, materials, buildings and people.

Work study was widely known for years as 'time and motion study', but with the development of the technique and its application to a very wide range of activities it was felt by many people that the older title was both too narrow and insufficiently descriptive. The term 'work study' entered the English language only after the Second World War, but it is now generally accepted; 'motion and time study' is however still used in the United States although the newer term is gaining currency there. The word Arbeitsstudium, which has a similar meaning, has been used in Germany for many years. (ILO, 1969)

The success of work study stems from the fact that it is a systematic way of investigating a problem and developing a solution.

Work study is composed of two groups of techniques:
- *Method study.*
- *Work measurement.*

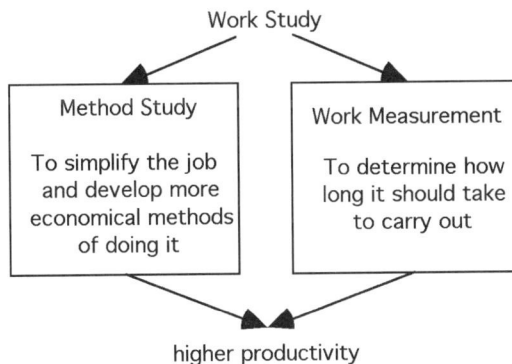

Figure 4/3.1 Components of work study.

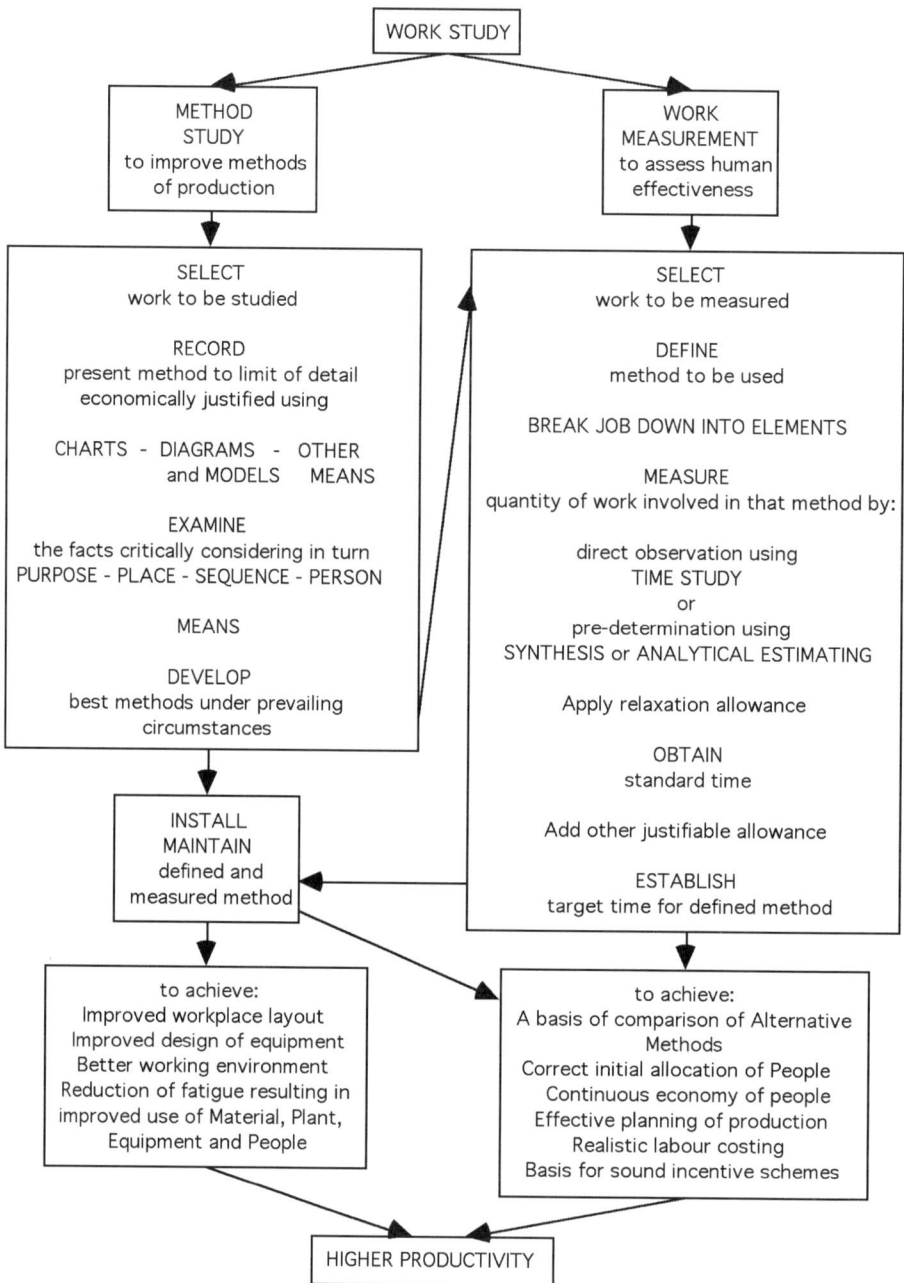

Figure 4/3.2 Method study and work measurement; coordinated procedures (after Currie, 1959).

(Fig. 4/3.1). The former is concerned with the way a work task is carried out while the latter is concerned with its measurement. Both techniques should be used jointly for maximum benefit. Their coordinated use follows Figure 4/3.2. Much of work study is formalised common sense, yet it is regarded as an essential tool for managers.

> *Work study acts like a surgeon's knife, laying bare the activities of a company and their functioning, good or bad, for all to see. There is nothing like it for 'showing up' people, and for this reason it must be handled, like the surgeon's knife, with skill and care. Nobody likes being shown up, and unless the work study specialist displays great tact in his handling of people he may arouse the animosity of management and workers alike, which will make it impossible for him to do his job properly.*
> (ILO, 1969)

A goal of work study is to improve what already exists. This implies change. The human context of this change is particularly important – how change affects all the people concerned, people at all levels of the team or organisation. Change can pose a threat to management.

> *... any proposed change must necessarily at first sight appear to be a criticism and hence in all probability a threat to the prestige or future prospects of the person con-cerned. It follows also that this kind of reaction will be the more intense the longer the service with the company and the more entrenched the man's position.*
>
> *Apart from this manifestation of the general conservatism of human beings, any change in existing methods and practices is an interruption of a comfortable situa-tion in which well established routines have been set up.*
>
> *Another possible source of difficulty is that when a work study specialist or team is active in a plant or office, it sometimes for the time being appears to take on a management role; so that the sitting managers are likely, unless the situation is handled with tact, to feel that their position has been usurped.* (Currie, 1959)

There is always a case for someone distant from the workface to objectively view the work procedures and make recommendations for change. The advantages of a systematic method study of work practices sanctioned by custom and established many years are emphasised.

The influence of work study on individuals can perhaps be ascertained from social research in the workplace:

> *Most researchers have observed that foremost among the expressed needs of people at work is money, ... once the reasonable financial needs are satisfied, other needs become more prominent. But these do not always emerge in the same order, and the main influences here seem to be connected with the position a person occupies at work. The hourly paid worker, given an adequate wage, wants security in that wage. The salaried worker, for whom a greater degree of job security exists, tends to go for increased status.* (Currie, 1959)

If the relationship between workers and management *...is already a bad one almost anything will strain it, because the workers will be suspicious of everything the management does. Equally, if the workers have confidence in the sincerity and integrity of the management, almost any sound technique will be accepted and can be made to work.*

In point of fact, work study, when properly applied, tends to improve industrial relations. (ILO, 1969)

There are several reasons why this should be so, namely:
* Involvement of the worker in the process.
* Worker kept informed.
* Work interruptions reduced, allowing the worker to get on with his/her job.

However,

(1) There may be strong resistance to changes in method proposed following method study, especially from older, skilled workers. Indeed, it may prove impossible to get some workers to change to new methods. If the methods they use and their output are reasonably satisfactory they will have to be left alone, changes being taught to new workers only.

(2) Many workers resent being timed: this may be due to either a suspicion of the stopwatch – which can usually be dispelled if its use is properly explained – or simply that having someone standing watching them worries them. Here the position which the work study man takes up and the way in which he goes about it is extremely important. He must take care to allow the worker to become accustomed to his presence before attempting to record times.

(3) There is often a fear that redundancy may arise out of the results of work study, leading either to unemployment or transfer to another department. (ILO, 1969)

In terms of the effect of work study on groups, this is perhaps a more delicate area:
Those responsible for introducing work study do well to remember that the work study specialist is a person from outside the working group which he studies and may therefore be regarded – until the appropriate explanations have been made – as a threat to the equilibrium of the group. From this it follows that, if the effect of work study is to break up or rearrange working groups, every effort must be made by patient, and perhaps even individual, consultation and detailed information to explain the reasons for the change and to gain acceptance for it.

All bodies of people who live and work together for a long time develop common traditions and established customs; but by its very nature work study is a challenge to tradition and custom, and seeks to replace for these aspects of work a closer discipline relating to the facts. (Currie, 1959)

Concern for all the people involved is paramount.

Quality, it is argued, should not suffer but rather improve through better work practices.

4/3.3 METHOD STUDY

Method study is concerned with finding better ways to do things. Desirably the cost savings from the method change should outweigh the cost of the method study.

> *... method study involves the breakdown of an operation (or procedure) into its component elements and their subsequent systematic analysis. Thence those elements which cannot withstand the tests of interrogation are eliminated or improved. In applying method study the governing considerations are, on the one hand, economy of operation and, on the other, the maintenance of accepted good practice as laid down by management (for example, safety and quality standards). (Currie, 1959)*

The goals of method study are:
- *The improvement of processes and procedures.*
- *The improvement of factory, shop and workplace layout and of the design of plant and equipment.*
- *Economy in human effort and the reduction of unnecessary fatigue.*
- *Improvement in the use of materials, machines and manpower.*
- *The development of a better physical working environment. (ILO, 1969)*

Technical matters and people should also be considered.

Method study might also be referred to as *motion study* in some texts.

The general procedure of method study is shown in Figure 4/3.2 and is broken up into the basic steps of:
- *Select*   (the work to be studied)
- *Record*   (all the relevant facts of the present or proposed method)
- *Examine*   (those facts critically and in sequence)
- *Develop*   (the most practical, economic and effective method, having due regard to all contingent circumstances)
- *Install*   (that method as standard practice)
- *Maintain*   (that practice by regular routine checks)

The steps are carried out systematically in order and all are necessary. Work measurement may need to be carried out in part to assist with obtaining information for these steps.

Some typical areas where method study may bring savings are:
- *Poor use of materials, labour or machine capacity, resulting in high scrap and reprocessing costs.*

- *Bad layout or operation planning, resulting in unnecessary movement of materials.*
- *Existence of bottlenecks.*
- *Inconsistencies in quality.*
- *Highly fatiguing work.*
- *Excessive overtime.*
- *Complaints from employees about their work without logical reasons.* (Currie, 1959)

This provides a hint to the selection process of the work task to be studied.

Recording

Results of recording tend to be displayed visually for easier analysis and 'before and after' comparisons as charts, diagrams or models or a combination of these. A number of recording techniques are used depending on the type of information that is recorded and the detail of the information. More than one technique may be used in any situation. The techniques are not ends in themselves but rather one step in the whole method study.

There are two categories of techniques:
- *Charts* (for process and time records).

Within these categories the most generally used techniques are:

Outline process chart	Principal operations and inspections.
Flow process chart	Activities of men, material or equipment.
Multiple activity chart	Activities of man and/or machines on a common time scale.

- *Diagrams and models* (for path or movement records).

Within these categories the most generally used techniques are:

Flow and string diagrams	Paths of movement of men, materials or equipment.
Two- and three-dimensional models	Layout of work-place or plant.

As well, older texts on work study mention two-handed process charts, simultaneous cycle (simo) charts, cyclegraphs and chronocyclegraphs. Two-dimensional models are sometimes termed templates. There are (in addition to multiple activity charts) *time charts* of the bar type which are used to represent when activities are being carried out or when resources are required. These are commonly called Gantt charts, bar charts or Gantt bar charts.

Time lapse recording and video recording are more recent additions to the armoury of method study.

Critical examination

The critical examination is the heart of method study.

This takes the form of a systematic analysis of the purpose, place, sequence, person and means involved at every stage of the operation, satisfactory answers being required in turn to each of the following questions:

(i)	(a) *What*	(is achieved)?	(b) *Why*	(is it necessary)?
(ii)	(a) *Where*	(is it done)?	(b) *Why*	(there)?
(iii)	(a) *When*	(is it done)?	(b) *Why*	(then)?
(iv)	(a) *By whom*	(is it done)?	(b) *Why*	(that person)?
(v)	(a) *How*	(is it done)?	(b) *Why*	(that way)?

A satisfactory answer to the query 'why?' leads in each case to consideration of alternatives which might also be acceptable, and finally to a decision having to be made as to which, if any, of these alternatives should apply.

Creative thinking and brainstorming-type techniques may be invoked as a means of generating alternatives.

New alternatives should take into account the human element including sight and lighting, colour, ventilation and heating, noise, seating and amenities.

4/3.4 WORK MEASUREMENT

Work measurement is given as:

The application of techniques designed to establish the time for a qualified worker to carry out a specified job at a defined level of performance.

The procedure is summarised in Figure 4/3.2.

Work measurement may be used in the following settings (ILO, 1969):

(a) *To compare the efficiency of alternative methods. Other conditions being equal, the method which takes the least time will be the best method.*

(b) *To balance the work of members of teams, in association with multiple activity charts, so that, as nearly as possible, each member has a task taking an equal time to perform.*

(c) *To determine, in association with man and machine multiple activity charts, the number of machines an operative can run.*

The time standards, once set, may then be used:

(d) *To provide information on which the planning and scheduling of production can be based, including the plant and labour requirements for carrying out the programme of work and the utilisation of available capacity.*

(e) *To provide information on which estimates for tenders, selling prices and delivery promises can be based.*

(f) *To set standards of machine utilisation and labour performance which can be used for any of the above purposes and as a basis for incentive schemes.*

(g) *To provide information for labour-cost control and to enable standard costs to be fixed and maintained.*

... work measurement provides the basic information necessary for all the activities of organising and controlling the work of an enterprise in which the time element plays a part.

The following are the principal techniques by which work measurement is carried out (ILO, 1969):

- *Time Study.*
- *Activity Sampling.*
- *Synthesis* from standard data.
- *Analytical Estimating.*
- *Production Study.*

With all techniques, tasks are broken down into their component or element times.

An element is a distinct part of a specified job selected for convenience of observation, measurement and analysis.

A work cycle is the sequence of elements which are required to perform a job or yield a unit of production. The sequence may sometimes include occasional elements. (ILO, 1969)

The selection of the work to be studied may be because of:

(1) *The job in question is a new one not previously carried out (new product, component, operation or set of activities).*

(2) *A change in material or method of working has been made and a new time standard is required.*

(3) *A complaint has been received from a worker or workers' representative about the time standard for an operation.*

(4) *A particular operation appears to be a 'bottleneck' holding up subsequent operations and possibly (through accumulations of work in process behind it) previous operations.*

(5) *Standard times are required prior to the introduction of an incentive scheme.*

(6) *To investigate the utilisation of a piece of plant the output of which is low, or which appears to be idle for an excessive time.*

(7) *As a preliminary to making a method study, or to compare the efficiency of two proposed methods.*

(8) *When the cost of a particular job appears to be excessive.*

If the purpose of the study is the setting of performance standards it should not normally be undertaken until method study has been used to establish and define the most satisfactory way of doing the job. (ILO, 1969)

4/3.5 RELATIONSHIP TO PROBLEM SOLVING

Systematic problem solving/systems engineering is carried out in the steps of:
- Defining the problem (including issues and situation).
- Selecting objectives.
- Generating ideas, alternatives.
- Analysing ideas, alternatives.
- Selecting the best alternatives (evaluation).
- Action.

The direct relationship with the method study half of Figure 4/3.2 can be seen. (The other half of Fig. 4/3.2 involving work measurement supports the method study half with data.) The objectives relate to maximising productivity, the generation of alternatives is carried out in the 'develop' step, and so on.

A direct relationship between value management and work study is also apparent.

4/3.6 REENGINEERING

Reengineering was originally perceived as a radical redesign of an organisation's work flows in order to decrease costs and time, and improve effectiveness; work study examines work in all its contexts leading to a systematic investigation of all the factors which affect the efficiency and economy of the work situation being reviewed in order to effect improvement. Reengineering has captured the minds of trendy management; work study grew out of the politically incorrect 'time and motion' studies in manufacturing, yet is equally applicable in the service industries and administrative and clerical fields. Many libraries have long since thrown out their 1950s books on work study because of lack of readership; these have been replaced by the tomes of the 1990s management gurus who appear not to be aware of work study methodology.

The paper 'Reengineering and Work Study' (Carmichael, 1997) argues that reengineering is no more than a case of 'reinventing the wheel' and couching it in appropriate buzz words and hype. As such, the questions are asked: what is new, and why do people embrace reengineering but regard work study as old fashioned and therefore irrelevant?

It is postulated that reengineering exists because the originators and users of reengineering were/are unaware of the body of knowledge of work study. Correct attribution should go to work study.

PART B

STARTING A PROJECT OFF

CHAPTER 5

Early Project Activities

A project has a beginning and proceeds generally in a logical fashion, but with some feedback loops, and always looking ahead. Starting a project off might be called a *concept phase, initiation phase, pre-investment phase*, or similar terminology. It involves activities of investigating the possibility of proceeding with a project and concluding with the project owner being satisfied that the end-product and project are feasible and being prepared to outlay expenditure for the project to proceed.

Generally it is the case that the decision making processes, involved in choosing which is the preferred proposal to invest in, have to be carried out in situations involving some uncertainties. These uncertainties may be acknowledged in a qualitative or quantitative fashion.

Typical activities involved in starting a project off include:

- *Establish a project's origin*
  [Origin of the project; The need for the project; Any market or opportunity studies undertaken or relationship to the organisation's strategic plan; Other origins; Data collection; ...]

- *Establish objectives and constraints*
  [Project objectives and constraints; End-product objectives and constraints; Origin of objectives and constraints; ...]

- *Establish scope*

- *Determine owner involvement*
  [In-house versus outsourcing/contract decisions (Carmichael, 2000); Joint ventures, ...]

- *Appoint project manager*
  [Selection and appointment; Qualities and characteristics, responsibilities; Project management delivery method (Carmichael, 2000); Project management services contracts; Legal matters, liabilities, ...]

- *Think about the potential project team*
  [Looking ahead to team building.]

- *Identify stakeholders*
  [Identification and management; Project owner; Project manager; Potential project team; Community; Community consultation; Authorities; ...]

- *Undertake support studies*
  [Data collection; Site inspection; Support studies – site/locale, market, risk/uncertainty and assessment, environmental (natural); Standards and regulations; ...]

- *Generate alternatives*
  [Generation of alternative end-products and means to end-products; ...]

- *Perform approximate estimating*
  [Approximate estimates of: resources, money (funding, income), times (schedules).]

- *Establish feasibility*
  [Technical and economic feasibility; Pre-feasibility studies; Feasibility studies; Risk analysis; Economic evaluation; Non-economic issues, ...]

- *Obtain approvals*
  [Summary report; At the end of the project initiation work, present a summary report of findings and recommendations. Approval to proceed further could be expected to depend on this; Authorisation; Obtain approval from – Statutory authorities, owner, ... Often the transition between project phases.]

Figure 5.1 gives an approximate flow of activities involved in starting a project off.

In the terminology of Chapter 16, there are two inverse systems problems here – one involving the end-product and one involving the means of getting to the end-product. The inverse problem on the left side of Figure 5.1 is usually referred to as the *design* problem or similar (Carmichael, 1981). The inverse problem on the right side of Figure 5.1 is one aspect of *project management* decision making.

The two problems are related, but practitioners flip flop between treating them as separate problems, and treating them as combined problems, particularly with regard to thinking on objectives and constraints. The distinction between the two problems becomes blurred. For example, the term 'deliverable' is commonly thrown around (and some people would say it is over-abused jargon) – its consensus meaning is that of end-product plus state (See Chapter 16) final (terminal) conditions for the project.

All activities in Figure 5.1 are legitimately part of what most people would regard as the project.

Management is an inverse systems problem (Chapter 16). As practised, suboptimal solutions are obtained generally because objectives are not defined beforehand, and hence the location of any proposed solution relative to the optimum is unknown.

Although drawn in sequence in Figure 5.1, many of the activities may be carried out in parallel. There can also be feedback between the activities as new information comes to hand. The order of the activities could be expected to be generally as listed.

Note that for some projects, all of these activities may not be present or may be only present in small amounts. Some projects place a significant emphasis on certain activities and disregard other activities. The work carried out is tailored to the particular project.

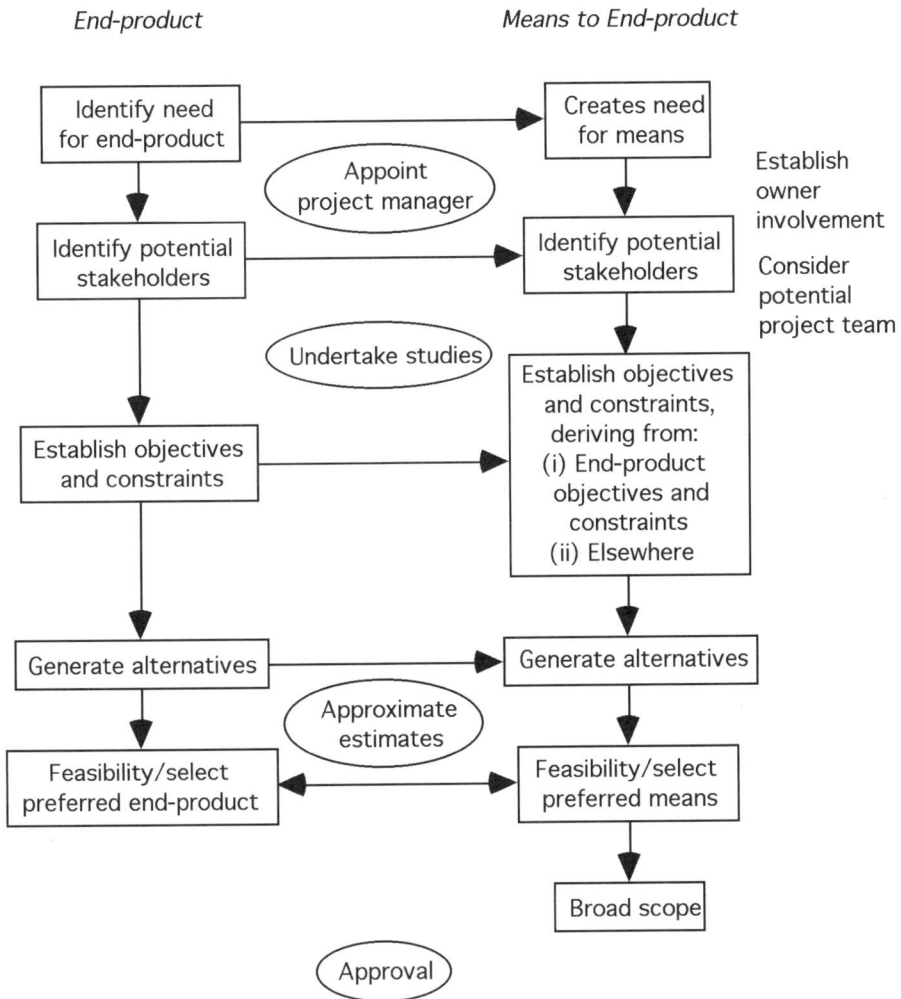

Figure 5.1. Project initiation activities.

In some projects there may be very little initiation phase work, most of it being done by others.

There may also be a blurring between activities carried out in different project phases.

Commonly, having reached the end of Figure 5.1, a proposal or report may be produced, and further project work may be dependent on obtaining approval to do so, from the owner.

The above and the following comments and notes are offered as reasonably common practice, but other versions can be found.

Case example

Road upgrade

Upgrading a road on a small island, to a sealed standard.

Need: The intention was to enhance the overall attractiveness of the island as a tourist destination, while assisting local industries.

End-product objectives and constraints included:
• Improved accessibility.
• Improved traffic safety; reduction in the accident rate.
• Improved travel times.
• Minimum whole of life costs.
• Enhanced natural environment; net positive environmental impacts.
• Improved access.
• Reduction in vehicle operating costs.
• Reduction in road maintenance.
• Performance criteria.

Stakeholders: The road ran through sensitive areas of native vegetation and a national park, and provided a significant tourist route as well as access for local communities including fishermen and farmers. There was existing conflict between local environmental (natural) groups, council and sections of the local community over previous and proposed roadwork. Stakeholder (including the wider community) consultation was recognised as a critical aspect of this project.

Scope: To seal 60 km of this road, and manage the associated activities.

Project objectives and constraints included (additional to those flowing from the end-product objectives and constraints):
• Estimated cost limit (budget).
• Cash flow and staging.
• Start date.

Examination of alternatives: Alternative solutions were considered based on technical, economic, environmental (natural) and community considerations.

Case example

Opera

The case study concerns the mounting of an opera by an opera company.

The sole business of the opera company is the production of opera. The opera company operates on a twelve month, seasonal basis. Over the course of its working year for example, the opera company will produce an average of 15 operas during its summer, autumn and winter seasons giving rise to about 225 performances. In addition, limited overseas touring is undertaken from time-to-time.

To accommodate such a schedule, 5 operas are usually in production and performance at any one time. As each opera finishes its run of performances (usually spread over 4 to 6 weeks) a new opera is opened to take its place. Hence, adherence to a rigid set of time schedules must be maintained if the overlap and parallel production of operas is to be met and sustained.

The artistic creative team – director, conductor, repetiteur (vocal coach and rehearsal pianist) and stage manager – work with principal singers and the chorus in a studio rehearsal situation on a full-sized mock-up of the 'set'. This moulding/shaping process takes about six weeks of production time before the cast is ready to combine with the orchestra for a sitzproben (seated) rehearsal and the final relocation to an opera theatre for 'stage' and 'dress' rehearsals prior to opening night.

While this activity is in rehearsal, the construction of the 'set' to be used on stage in performance is close to completion. Depending on the complexity of the set design, set construction under the watchful eye of the technical director and his/her heads of departments can take up to four months. Moreover, as the company has its own internal and highly skilled manufacturing operation, many occupational health and safety (OH&S) issues relating to the building and workability of the set are addressed and monitored both for the crew during construction and the artists while in performance.

In parallel with these activities, the wardrobe department must acquire the material resources to manufacture, in accordance with the costume designer's detailed drawings, the costumes, wigs, shoes and accessories for each cast member of each opera.

Added to this intense period of activity is the orchestra's rehearsing of the score, the marketing and communication departments' promotion of the opera through their advertising networks and media connections and the development department's endeavours to secure sponsorship for this and other operas performed during the season.

Critical to the mounting of opera therefore, is the timing of these independent activities undertaken by separate departments who are at times oblivious to what phases other departments are at or on what opera they are actually working. See Figure 5.2. Each administrator is responsible for completing his/her component of the project on time and to internationally benchmarked performance standards. The coordination of departments and the comprehensive scheduling of the company's activities to meet strict deadlines is the responsibility of the artistic director.

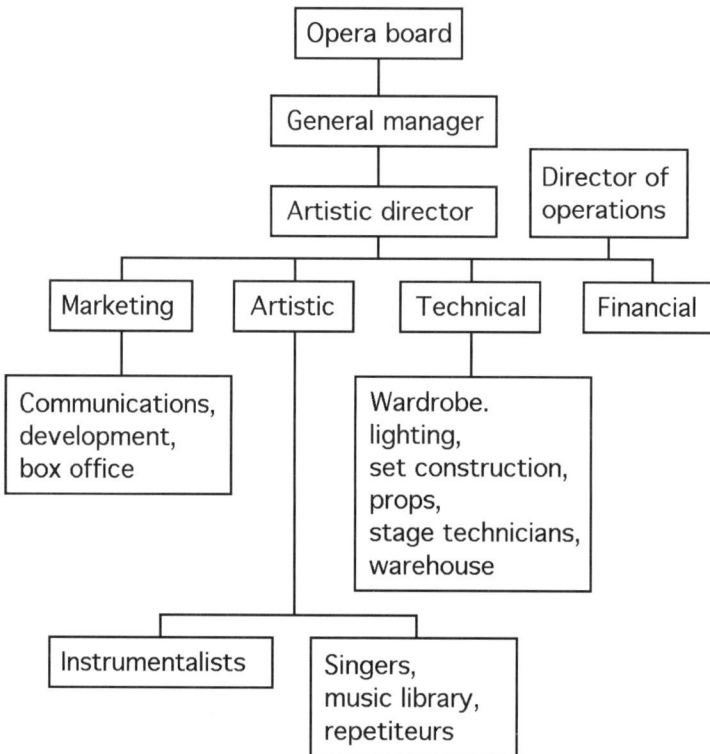

Figure 5.2 Management structure of the opera company.

The decision as to which combination of operas to produce in any one season also rests with the artistic director. These decisions are usually formalised 5 years in advance to allow enough time to secure the artistic and technical creative teams, who the artistic director believes will bring inspiration and a new interpretation to productions that audiences may have seen many times before. Moreover, because the marketplace for international and national artists (singers and conductors) is highly competitive, a similar lead time for contracting their services is also required.

When all arrangements have been finalised, a seasonal plan and cast list is prepared and presented to the opera board for approval. In line with the company's corporate and strategic plan, a mix of traditional and contemporary operas together with a balance of new, premier productions and revivals of existing productions (a recurring four year cycle until replaced by a new production) places great pressure on the cultivation and securing of corporate sponsorship to cover the shortfall between government funding/box office return and actual operating expenditure. The role of the general manager is therefore to oversee the business operations of the company, its positioning in the industry, its corporate image and appeal and, as a major cultural organisation, secure on-going financial support from government.

Identifying needs

As project manager, the artistic director is charged with identifying audience needs and creating an operatic season that is culturally satisfying and challenging. To achieve this, the artistic director, in conjunction with the marketing department, implements an historical review and analysis of previous productions from both a financial and artistic viewpoint. Other ways and means of identifying audience preferences come from subscriber's correspondence, survey forms distributed via seat drops in theatres and formal questionnaires undertaken by a marketing research company. Formal investigations are usually undertaken every 5 to 10 years and provide valuable information and assistance in the decision making process. However, the artistic director, while taking this collated data into consideration, still maintains overall control by weaving his/her own choices into a season's format.

Objectives and constraints

The artistic director's job is to not only meet the company's mission and corporate goals, but promote artistic values, strive for improved performance standards and challenge audiences with new and confronting opera productions that broaden their horizons. The artistic director is regularly faced with financial constraints that may require his/her asking for an alternative set design to the original concept, or his/her re-casting of roles so as to fall within budget limitations, or a re-vamp of a previous costume wardrobe rather than manufacturing specially created costumes based on new designs. These dramatic changes and enforced decisions are often the result of fluctuations in projected audience figures, or box office estimates and sponsorship income or the sudden unavailability of a key singer or all three and more, and underpin the reasons why there is a blow-out in the company's bottom line: primarily from the artistic director's unwillingness to compromise artistic standards just to balance the books. Hence, as the intangible benefits far outweigh the tangible, most leading opera companies continue to operate on a healthy deficit.

Risks

Anticipation of unforeseen events that occur to singers for example, requires the coaching of understudies who are prepared to stand in at short notice in a performance. The box office is another high risk area where errors in booking and seating allocations often occur. Computerised links to other points of sale also cause unexpected problems when the system 'goes down'. Companies which revise their sponsorship commitment can create unforeseen financial problems that require urgent attention. Accordingly, entire production costs are covered in-house from time-to-time because sponsorship could not be found. Moreover, no matter what contingency plans are devised, not all are successful. However, once the production is underway in all sectors of the company, little change in scope occurs.

Feasibility

The artistic director, besides reviewing previous productions and implementing revivals every four years, visits prominent opera houses to preview new, quality productions. His/her decisions regarding suitability for staging and meeting local audience expectations are presented for approval to the company's board. If a production is to be added to the opera company's repertoire, further decisions such as hiring and transporting the complete production versus the recreation of the production locally using the original creative team, have to be made.

Cost and time constraints are two primary areas to be identified and monitored if such a project is approved. The availability of top-line singers for an extended period and the procuring of a suitable conductor demands a minimum lead time of five years if the best affordable performance team is to be assembled.

Work breakdown

Sets
From a costing carried out on pictorial set designs and models, the procurement of materials is investigated and materials acquired on a just-in-time (JIT) basis. Modifications to the set due to cost constraints, reduced stage area or source materials is requested of the set designer. Each component of the set is subsequently identified as a subproject and allocated to specialist teams of skilled workers who convert the pictorial designs into plans (design workshop), construct (carpenters, steel fabricators, engineers for hydraulic systems, fibreglass fabricators) and finally paint (scenery artists) their component of the jigsaw puzzle.

Wardrobe
A similar breakdown occurs in wardrobe by buyers, the cobbler, wig makers, costume seamstresses, milliners, jewellery and accessory makers and make-up artists.

Artistic
See Figure 5.3.
   In the artistic sector, the chorus is trained separately by the chorus master, working on two or three new operas additional to those in performance. From the chorus, a senior member is usually selected as an understudy and offered a suitable covering role. Repetiteurs (staff pianists) individually coach these singers. The conductor supervises ensemble rehearsals and provides details on performance and musical direction. Each repetiteur is allocated particular operas to musically prepare, to prompt in the prompt box and coach roles, as well as being required to speak several languages fluently. In production rehearsal calls, repetiteurs play the piano in lieu of the orchestra, while the director instructs the artists and chorus in stage direction. A movement specialist is engaged at times to assist the director and singers, and reduce potential repetitive strain injuries and occupational health and safety (OH&S) problems.

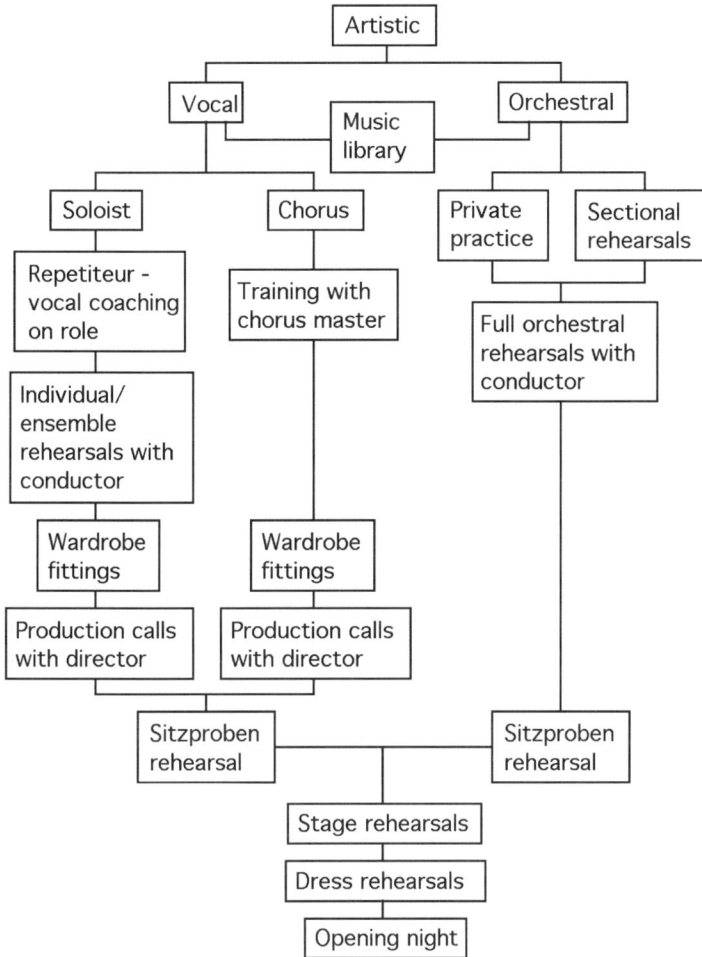

Figure 5.3 Decomposition of artistic activities.

Very little overlap occurs between departments, except for scheduled costume fittings or when production calls have concluded and principal singers and chorus come together with the orchestra for the first time. The orchestra has, up to this point in the development phase, maintained separate its performance and rehearsal schedule, preparing for a new opera, while individual players can borrow their part from the music library. In a more formal way, section leaders have the ability to call sectional rehearsals if necessary, while the conductor can formally allocate a scheduled rehearsal call to a particular section of the orchestra and ask for an additional tutti rehearsal for another time. Budgetary and award constraints may mean his/her requests have to be delicately declined.

A sitzproben is the first joint rehearsal for singers and orchestra prior to moving the production to the opera stage proper. With repetiteur close at hand taking notes, the conductor can refine performance standards even further. At the end of this long development phase, 'stage' rehearsals draw all facets of the production together for the first time. Working on the set in costume under lighting in the performance venue with the orchestra in the pit is fraught with potential mishaps. The artistic director in overseeing all adjustments, carefully balances the temperaments of his/her creative production team and is ready to appease for the sake of meeting the production's opening night deadline. After two or three stage rehearsals have been completed, next comes a dress rehearsal or a complete run of the opera in performance so that singers can pace themselves and any technical glitches that occur be rectified before the first performance.

Success or failure

The opera company's approach to the mounting of an opera, once the initiation of a season has begun five years ahead of time, relies on historical records, archival recordings, past practices and streamlined efficiencies. The company's area of highest expenditure is salaries. Balancing international artists and artisans with a local full-time workforce helps reduce these costs. However, because the artistic director pays little heed to the financial repercussions of his/her opera selections and casting decisions, the company's operations rarely reach a break even point. His/her argument is that if local singers of less stature and ability than their overseas counterparts were engaged, a reduction in audience numbers would result. To balance the mix of drawcard singers and local product with exciting, young, as yet little known designers and conductors with sort-after directors is often difficult, and when out of balance, box office suffers. Another factor in estimating costs is the number of performances for each opera. In this area the artistic director's feasibility studies of opera preferences and for some operas, the number of allocated performances, can be questioned.

The company's approach to an opera is very detailed and highly productive with the resultant quality admired widely within the industry but by a minority of the general public. Other commercial enterprises in the industry who mount musical stage productions never reach the same levels of performance standard or quality control in any area as the opera company does but have little trouble securing sponsors or attracting an audience for a considerable number of performances, purely because they found the ingredient that captures the imagination of the general public.

The approach does not disclose to the artistic director the facts as to why one production is highly successful and another with the same quality of preparation, a box office flop. Certainly, the quality of craftsmanship and the timely conclusion of an opera's preparation with published performance dates is highly prized. Moreover, the ability to prepare and mount operas in parallel while individual activities constantly overlap indicates an efficient project management process. But, aside from all these positives, what it finally comes down to is the intangible benefits outweighing the tangible; this being the main element which determines the success or failure of an opera.

The success of the opera company's operation is its in-house manufacturing and training systems. All production activities associated with the technical, artistic, marketing and financial departments are coordinated within the project management process to achieve the end-product. The subdivision of an opera into major subprojects such as the planning phase, the development (manufacturing) phase and implementation phase or performances is no different to most other projects. The renown of the artists, who are engaged on the project, provide the catalyst for the marketing and development departments. The product's worth is far easier to sell for example if an internationally known artist stars in the production.

The decomposing of the project into separate departments to facilitate control over what is to be delivered on reconstitution, regular meetings of senior administrators with the general manager to report on the progress of activities, and monthly financial reviews to keep a watchful eye on expenditure, so that alternatives to reduce the impact of budget blow-outs can be instigated, identify the opera company as a leader in its field, but it is felt raises doubts over its competitive ability to streamline project efficiency and productivity when no other competitors in the area of opera production exist.

EXERCISES

1. Given that the early work done in a project has the most ability to influence the cost of a project, why is it then that examples continually occur of projects proceeding without thorough preparatory work?

Is there a false economy here of project owners saving money initially in the belief that the total project cost will consequently be less?

How true is the Confucian saying:

In all matters success depends on preparation.
Without preparation there will always be failure.

The work done in the initiation phase is the 'preparation'.

Give an example, from your workplace, where something went wrong in a project because no preparatory work was done.

2. The term 'deliverables' is only a recent introduction to project management. What term was used before its introduction?

CHAPTER 6

A Project's Origin

6.1 INTRODUCTION

The origin of any project is the recognition/identification that a need, a want, a new or expanded market, or an opportunity exists. There is a reason for undertaking the project as a means of getting to an end-product (product, facility, asset, service, ...). The source may be variously:
- An organisation's strategic plan.
- Market research.
- Opportunity studies.
- Prior obsolescence.
- Political input.
- Tendering.
- etc.

and combinations of these.

6.2 STRATEGIC PLAN

For many public-sector and private-sector organisations, the recognition of the need or desire for some product may flow directly or indirectly from that organisation's strategic (long-term) planning.

The strategic (corporate, business) plan has incorporated into it the organisation's *mission* (a description of the organisation's business). Associated with a mission is a *vision* statement embodying a longer term direction for the organisation. Both ideas of mission and vision may be regarded by many as faddish management expressions and practices.

The corporate plan contains broad aims which might be called goals, key result areas or similar. [The term 'objective' is deliberately not used here. See Chapter 7.] The corporate plan is the basis for an organisation's direction and priorities. These aims are passed down the organisation levels/hierarchy becoming more definite as the level drops. Planning in this manner is a top-down approach.

(Some strategic planning goes bottom up before it becomes top down. That is, lower levels of an organisation contribute, but consolidation then comes from the top.)

For a research and development oriented company, the company's corporate plan would be expected to indicate how much is to be spent on research and development and what key areas of research are to be focused on for continued growth. Project selection becomes a critical and very difficult decision area. Apart from corporate criteria, projects are expected to meet certain marketing, research and development, financial and production requirements. Any technological breakthrough would be expected to be matched by a reasonable expectation of commercial success in order to get project approval.

The project origin may thus be presented as *fait accompli*, in the form of a brief or similar. The brief defines the parameters within which the project and the end-product must lie. Much of the project initiation work, such as market research, generating alternatives, and feasibility studies is unnecessary. Broad identification, feasibility, ranking and prioritising will have been carried out at the higher levels of the organisation.

Case example

Company Business Plan for Year ...

Our Mission
We will be the undisputed worldwide leader in ... with solutions that ...

Our Values
Integrity; Innovation; Continuous Improvement; Performance; etc.

Our Vision
Satisfied Customers; Growth; etc.

Business and Customer Requirements
• Customer Focus.
• Responsiveness.
• Market Leadership.
• Innovative Solutions.
• Value for Money.
• etc.

Our Strategies for Year ...
• Develop and Publicise an Image.
• Improve Responsiveness to Market Requirements.
• Improve Sales, Engineering and Service Productivity.
• etc.

Our Measures, Primary and Secondary
• Customer Survey.
• Gross Margin %.
• etc.

Our Targets, Year ... and Year ...
- 10% increase in scores in survey per annum.
- 10% error decrease per annum.
- 20% gross margin.
- etc.

Case example

Road transport authority

Mission statement
To manage the use, maintenance and enhancement of the roads and traffic system with emphasis on road safety and transport efficiency as part of an integrated and balanced transport system.

Strategic plan
Refer Figure 6.1. The corporate plan contains broad goals for enhancing road systems which are part of an integrated land use and multi-modal transport system.

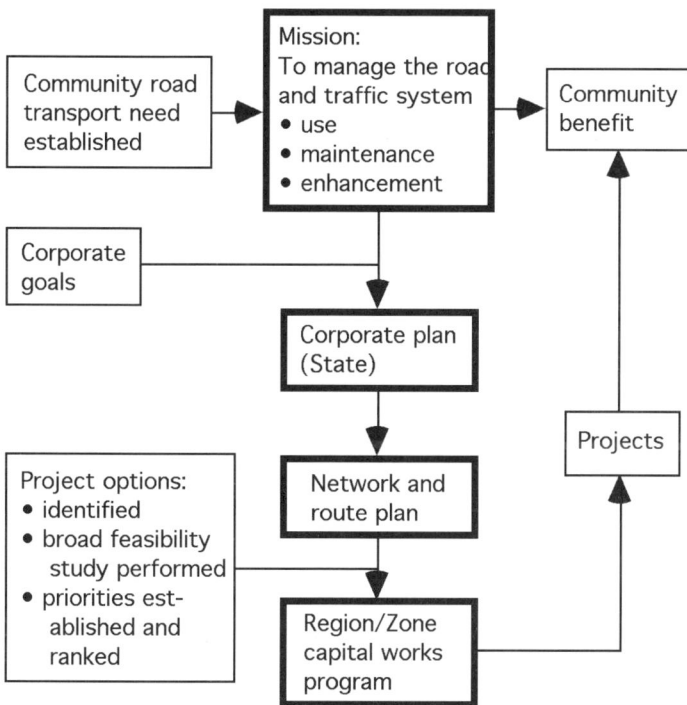

Figure 6.1 Strategic thinking and projects.

Key result areas (KRAs)
Natural environment
- Noise.
- Landscaping.
Efficiency
- Cost and timeliness of work.
- Quality assurance.
Transport efficiency
- Feasibility.
- Ride quality.

Corporate plan
The plan describes the authority's direction and future priorities after consulting internal and external stakeholders.

The corporate budget included in the corporate plan classifies expenditure on the basis of programs relating to capital works. The use, maintenance or enhancement of the roads and traffic system is based on planning priorities.

Network and Route plan
The plan provides medium and long-term strategies for developing transport routes based on growth and funding.

Plans for regional networks, both urban and rural, are prepared so as to be consistent with the broad authority goals. Specifically, these network plans meet the transport requirements necessary to support population growth and economic development.

Each project team must be aware of where its project fits in with network and route development goals and strategies. When developing and evaluating project options, the broader goals must be considered. Individual projects must be initiated and implemented so that network and route goals are best achieved (Fig. 6.2).

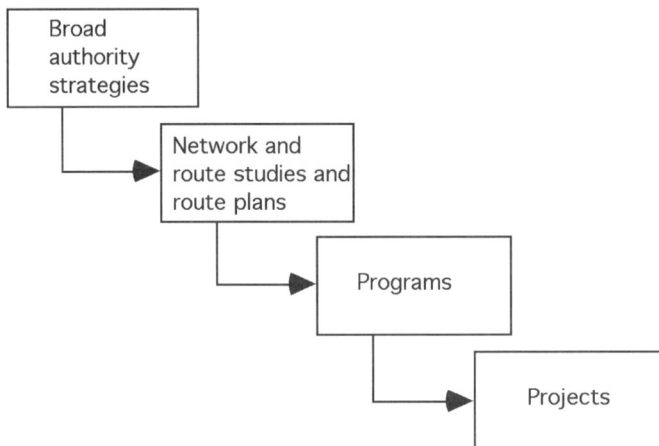

Figure 6.2 Network planning context.

Region and zone business plans
The plans provide a localised framework.

Region and zone business plans provide a local interpretation of the corporate plan. They identify local KRAs which reflect corporate aims and also identify priorities.

Regional business plans provide the framework for capital works programs, and are used to derive project objectives and performance indicators.

[based on RTA (1992)]

Organisation budgets

A similar origin is exampled by the annual compilation of the following year's 'budget' of an organisation. The various middle managers put together a list of items, their associated costs, justification of why they are necessary, and when in the financial year they are required. This is then overviewed by senior management with the view to seeing that the items fit within the overall strategic direction of the organisation.

In publicly listed companies, the process may go beyond, to the board of directors and shareholders for approval.

Through the year, the situation may change, necessitating a change in the budget and a change in the listed items. For example, new technology appears, and exchange rates may also change. Provided such changes are not gross, the 'budget' may have sufficient flexibility to change without seeking board of directors and shareholder approval. The 'budget' is continuously monitored and reviewed throughout the year.

At the end of the financial year, a reconciliation is undertaken of expenditure versus budget, and a review is made as to the worth of the expenditure.

Case example

Software introduction

This case study relates to the introduction of a new software product (management system) into the industrial process control marketplace.

The context behind the origins of the software were:
• The company sought competitive advantage through two main strategies – differentiated levels of customer satisfaction, and striving to be a solutions provider.
• The changing and evolving requirements of the process control marketplace.
• The gaps and shortcomings, which somewhat weakened the saleability and credibility of other software, in the company's existing product portfolio.
• The need to achieve revenue growth by extending the company's presence in plant control and information solutions.

All pointed to the area of management systems. In some respects, the company's current product deficiency had not been too much of an issue, as most of the company's major customers were not yet requiring solutions in this area. The company's primary offering at the time was a third party product, for which the company had a non-exclusive distributorship.

Over recent years, industrial manufacturers had changed from doing 'more' to doing 'smarter'. This trend was confirmed by many high level discussions, workshops and benchmarking studies done with the most successful of the company's customers. The studies were not done specifically as part of a need identification process, but the data was still useful. Ad hoc market research information existed on buyer analysis/segmentation, and the company possessed considerable insight into current and emerging market conditions. Data from third party industry 'think tanks' was useful in substantiating a view of the way the marketplace was going.

The shift stemmed from increased market and economic demands, such as competition, deregulation, and more stringent environmental (natural) regulations, affecting many industrial producers.

Thus, it became obvious to the company that the area of management systems was fast becoming the 'next big thing', as an enabler for manufacturing. In other words, it was a necessary component of the company's software philosophy, but was the area which was possibly most lacking in the existing product portfolio. The company did have a number of different offerings here, but in general there was not a single, integrated solution that enjoyed a competitive advantage. This was reflected in the fact that the company had lost some large tenders to competitors for management systems on existing customer sites.

What followed next was a systematic evaluation of various other management system solutions in the marketplace, including the company's own products. This evaluation looked at each product's functionality, cost, performance, technology, customer acceptance, market share and support requirements. The investigation of alternative software products was structured, with extensive primary data collected. A detailed evaluation matrix was developed, which compared a number of alternatives (including the company's existing products, and the current market 'leader') against many categories. Each category had an importance or weighting assigned to it, with relative scores recorded for each of the products. This evaluation was not completely objective, because it was compiled in-house, albeit by people very familiar with all the products being evaluated.

The feasibility of developing a new product from scratch was also investigated.

It turned out that a small overseas organisation had produced a world-class management system product, albeit with a relatively small but growing market share. The overseas organisation was targeting the same vertical markets, and thus had significant market credibility. The only major impediment to this organisation's run-away market success was the lack of a global sales/distribution channel, an area in which the company excelled. So, as the saying goes, 'we liked the product so much we bought the company'. Thus, a new product was 'born', albeit significantly developed and enhanced by the company's software engineering group to make it truly a company product.

The company had the choice of continuing with its existing products, or to adopt the new product as being the solution of choice. With progressive customers demanding new management system solutions, this meant that it was almost a given that the new product was going to be introduced.

Interestingly, the company found that the software opened up opportunities in several new markets that the company historically had not pursued. This came about partially because the software solution that the company implemented was quite generic, and could be implemented in many markets without the requirement for significant modification. This possibility was not considered in the original product requirement definition process.

Case example

The development of business banking facilities

Background

In a financial institution, having evolved from a mutual organisation which was competing in a corporate environment, projects traditionally resulted from strategic management directives to meet the needs of competitive market forces, industry requirements or government regulations. These projects invariably required substantial information technology resources.

Following corporatisation, there was a hectic pace of change within the organisation to meet the needs of its traditional customers, whilst at the same time satisfying the expectations of stock market analysts and the governing regulatory body. It is against this background that one project unfolded.

Business banking

As a mutual organisation that existed for the benefit of its members, the institution had never embraced business customers, because it essentially catered for the needs of individuals. Corporatisation, however brought new pressures and the expectation that it would perform against the same benchmarks as its competitors. To this end, an unwritten condition of its licence to operate as a corporate body, was the requirement to develop products that would satisfy the banking needs of small- to medium-sized businesses. That is, business banking was to be developed.

6.3 MARKET RESEARCH

Market research involves the search for and study of facts and information relevant to the owner's situation. It is intended to be done systematically though, in practice, this may not be the case. The activity of market research may be viewed and conducted as a project in its own right.

Components of market research include:
- Situation analysis.
- Formal investigation.
- Analysis and interpretation of data.
- Reporting.

The *situation analysis* involves obtaining information on the owner, and getting a 'feel' for the situation. Involved in this may be an analysis for the owner's business, competition and the industry in general.

The *formal investigation* extends any investigatory work done in the situation analysis phase. It principally involves:
- Selecting the sources of information.
- Selecting the methods of gathering data.
- Planning any sampling and forms.
- Collecting data.

Sources of information are varied but may be classified as:
- Primary data.
- Secondary data.

Primary data is data gathered specifically for the situation at hand. *Secondary data* is data gathered for some other purpose but still useful for the situation at hand.

Sources of secondary data include libraries, government publications, trade, professional and business associations, companies, and higher education organisations.

Primary data may be gathered by survey (personal interviews, telephone surveys, interviewing by mail), observation, or experiment (forms, questionnaires, sampling).

Particular issues addressed in market research include forecasting, buyer analysis/segmentation, choice process, factor choice/testing and product research.

Case example

School projects

Market research was conducted as part of strategic planning at a private school. The research was carried out so as to understand current attitudes towards the school held by the school's community. The results led to the initiation of several projects.

Those surveyed included parents of current students, and parents of children enrolled to attend but who, having paid an enrolment deposit, subsequently sent their children to other schools.

Those people who sent their children elsewhere
From the market research into this group a better understanding of perceptions was gained about:

- Who the school competitors were.
- The strengths of the school which attracted this group to enrol in the first place.
- The important role transport played in influencing the choice of school, especially in relation to students enrolled in the primary school.
- The school's suspicions about the influence of the poor reputation the local neighbourhood had.
- The school's belief that the small grounds and lack of open space had on the perception of others about the school.
- The influence the school's fee levels had on potential enrolments.
- The enrolment procedure and parent's perceptions about this procedure.

Parents of current students
From the market research into this group a better understanding of perceptions was gained about:
- The school's strengths and weaknesses.
- Major projects related to facilities which were perceived as highest priority.
- The role technology played in their children's learning experiences.
- The number, variety and quality of the subjects that the school offered.
- The physical facilities at the school.
- Modes of transport and travel times.
- The school's standing compared to their perceptions of other schools.
- Who the school's competitors were.
- The quality, nature, and sufficiency of information flowing from the school.
- Priorities for future spending.

Outcomes
A number of outcomes related to the information received resulted in variations in the scope and priorities of projects which formed part of the school's strategic plan, as well as the initiation of a number of smaller projects specifically in the area of information technology. The school also looked very carefully at aspects of its enrolment process.

The school believed that the market research was a valuable tool which helped it to better understand the market in which it functioned and the concerns and priorities of its customers. The school incorporated the results of the research into its strategic planning and rescheduled those projects which its customers perceived to be of the highest priority.

Market research as a project
The market research was a project in its own right (or a subproject of the development of the organisation's strategic plan). The designated project manager was formally allocated time for the work required, and needed a variety of skills to juggle the competing demands made during the course of the project, in particular the conflicts between normal day-to-day activities and those required to complete the project.

6.4 OPPORTUNITY STUDIES

Opportunity studies (when appropriate) are usually the first studies undertaken by an owner to identify investment opportunities. In many cases, strategic planning practices, market research and opportunity studies overlap. Opportunity studies can be either classified as *general* opportunity studies in which studies are undertaken in selected industry sectors such as water treatment, and mineral processing, or *specific* opportunity studies which simply investigate a specific investment proposal. General opportunity studies can be either *area, sub-sectoral* or *resource-based* studies. Figure 6.3 provides outlines of the various items included in the three types of opportunity studies for the particular case of chemical plants. In some instances, the extent of opportunity studies is such that pre-feasibility studies are dispensable.

A Outline of an area study

1. *The basic features of the area: the area size and leading physical features, with maps showing the main characteristics.*
2. *Population, occupational pattern,* per capita *income, and socioeconomic background of the area, all set in the context of the country's socioeconomic structure, highlighting differences of the area compared.*
3. *Leading exports from and imports to the area.*
4. *Basic exploited and potentially exploitable production factors.*
5. *Structure of any existing manufacturing industry utilising local resources.*
6. *Infrastructural facilities, especially of transport and power, conducive to development of industries.*
7. *A comprehensive check-list of industries that can be developed on the basis of the available resources and infrastructural facilities.*
8. *(Refines item 7.)*
9. *Estimation of present demand and identification of opportunity for development based on other studies or secondary data, such as trade statistics for the list of industries.*
10. *Identification (by considering the best economic sizes of plants and transportation costs) of the approximate capacities of new or expanded units that could be developed.*
11. *Estimated capital costs of selected industries.*
12. *Major input requirements. For each project approximate quantities of essential inputs should be estimated, so as to obtain the total input requirements. Sources of inputs should be stated and classified (that is, local, shipped from other areas of the country, or imported).*
13. *Estimated production costs to be derived from item 12.*
14. *Estimated annual sales revenue.*
15. *Organisational and management aspects of project sponsor(s), or a potential enterprise.*
16. *An indicative time-schedule for implementation.*
17. *Total investment contemplated in projects and peripheral activities, such as development of infrastructure.*

18. *Projected and recommended sources of finance (estimated).*
19. *Estimated foreign exchange requirements and earnings (including savings).*
20. *Financial evaluation: approximate pay-off period, approximate rate of return. Assessment of possible enlargement of product-mix, increased profitability and other advantages of diversification (if applicable).*
21. *A tentative analysis of overall economic benefits, and especially those related to national economic [goals], such as balanced dispersal of economic activity, estimated saving of foreign exchange, estimated generation of employment opportunities, and economic diversification.*
    Indicative figures based on reference programming data, performance of other similar industrial establishments should be sufficient for this purpose.

Figure 6.3a Outline of an area study (UNIDO, 1978).

B *Outline of a sub-sector opportunity study*

1. *The place and role of the sub-sector in industry.*
2. *The size, structure and growth rate of the sub-sector.*
3. *The present size and rates of growth of demand of items that are not imported and of those wholly or partially imported.*
4. *Rough projections of demand for each item.*
5. *Identification of the items in short supply that have growth and/or export potential.*
6. *A broad survey of the raw materials indigenously available.*
7. *Identification of opportunities for development based on headings 2, 5 and 6, and other important factors, such as transport costs, and available or potentially available infrastructure.*

Figure 6.3b Outline of a sub-sector opportunity study (UNIDO, 1978).

C *Outline of resource-based opportunity studies*

1. *The characteristics of the resource, the prospected and proven reserves, the past rate of growth and the potential for future growth.*
2. *The role of the resource in the national economy, its utilisation, demand in the country and exports.*
3. *The industries presently based on the resources, their structure and growth, capital employed and manpower engaged, productivity and performance criteria, future plans and prospects of growth.*
4. *Major constraints and conditions in the growth of industries based on the resource.*
5. *Estimated growth in demand and prospects of export of items that could utilise the resource.*
6. *Identification of investment opportunities based on items 3, 4 and 5.*

Figure 6.3c Outline of resource-based opportunity studies (UNIDO, 1978).

Case example

Medical offices

A company providing X-ray, radiography, mammography and nuclear medicine screening services in a metropolitan area considered three options – amalgamation, relocation and starting practices.

Whether amalgamation of, relocation of, or starting new practices takes place, this is based on research, and recognition of expanding suburban areas, taking into account:

- A suburban centre's regional and local growth.
- Doctor referral patterns.
- Availability of suitable premises – size, easily identifiable location, ground floor position, long lease (5 to 15 years).

Shareholders

In the private-sector, new projects stem from shareholders' expectations of a company's business and financial performance. Shareholders expect and demand higher and higher rates of return, forcing the company to continually and aggressively look for new markets and opportunities to do business in.

Case example

Timber importation

A company was looking at the opportunity of supplying the local market with imported timber. It was discovered, during market research, that there was a great potential in such a business because of the following reasons:
- The local building and construction industry was, to a large extent, dependant on timber products imported from overseas and there would be a demand for imported timber in the future.
- The product which the company was offering had a better texture than its local rival, and this would attract customers.
- The cost of acquisition, even including transportation costs, was relatively low due to low overseas labour costs.
- The competition in importing such timber was essentially nonexistent.

To learn this, the company had to undertake market research on the product it was offering and assess the opportunities for successful product launching on the local market.

The market research comprised personal interviews with industry and government representatives, customs officials and agents. The answers were sought on questions such as: 'Is there a need for the product?', 'What should the product and its major features be like?', and 'What competition would it face?' Compliance with government regulations and customs' requirements was also considered. Publications from the timber industry were studied.

The opportunity study looked at:
- The market and areas of competition.
- The issues of quality, packaging and customer requirements.
- The required drying and finishing of the timber.
- The most economical and effective way of packaging for transportation.
- The approximate competitive prices.
- Present and future activity in the building and construction industry.
- Government requirements for the industry; existing regulations.
- The procedures for obtaining approvals for treated timber; treatment of timber.
- The procedure of certification.
- Professional ethics existing in the industry.
- The issues of quarantine conditions.
- Customs duties.
- The transportation of timber in containers.
- The classification (grading) of timber products according to ability to withstand loads etc.
- Estimation of present demand and rough projections of demand for each item.
- Estimated production costs.
- An indicative time-schedule for implementation.
- Projected and recommended sources of finance.
- Major constraints and conditions in the growth of industries based on the resource.

The market research and opportunity study answered the question of feasibility and outlined the concept to be adopted for the business. The research indicated that the project could be successful provided that compliance with all the requirements and recommendations outlined in the study were unconditionally implemented. The area of competition was chosen along with the range of products and their proposed quantities and properties.

The market research was treated as a project. The marketing problem was well defined to ensure that time, money and energy were not expended unnecessarily in the research process.

Case example

Water/wastewater

Background

An engineering company, which had many years of experience with many success-ful projects in the milling and mineral processing industry, was actively seeking to diversify its potential market for its recognised EPCM (engineering, procurement, construction and management delivery method) services. This led to the strategic acquisition of another project engineering company with significant experience in the domestic and international water and wastewater markets. The acquired compa-ny's most recent water/wastewater projects were located overseas and its visibility and presence was poor in the domestic market.

Table 6.1 gives the company's assessment of the differences encountered between mineral processing projects and water/wastewater projects as they relate to the com-pany.

Table 6.1 Mineral processing projects compared with water/wastewater projects.

Mineral processing projects	Water/wastewater projects
Project is driven by economic factors.	Project is driven by environmental (natural) factors.
Delivery focus. Short delivery time desired by private-sector owner.	Project delays enable delay of forced investment.
Highly developed in-house expertise.	Restricted technological edge.
Company well known – market leader.	Company relatively unknown; may have problems prequalifying.
Most reference projects domestically.	Most recent reference projects overseas.
Typical project value large.	Typical project value small.
Turn key, D&C (design-and-construct) or EPCM reimbursable delivery.	Head contractor; design consultant.
Project delivery guarantee.	Delivery and performance guarantee mandatory.
Commodity price driven – (fluctuating workload).	Ongoing business – potentially increasing environmental (natural) workload in the future.

Need for the project

A high profile project was required to strengthen the local credibility for the com-pany's water/wastewater capability. A business plan was developed to ensure the future performance of this small, but growing business segment. The intent of the project was to:

- Generate short-term and long-term profit for the company.
- Feature innovative, up-to-date technologies and methods.
- Be a showcase for D&C, and EPCM capability.
- Develop and diversify the company's internal staff expertise in new technical areas.
- Showcase for international visitors (close proximity to the company's head office).
- Assess the company's competitiveness in the domestic water/environmental (natural) market.
- Display broad environmental (natural) capability to existing customers in the mineral industry.

The potential market

Potential projects were typically found in work conducted for public utilities. This type of work was typically advertised by public registration of interest, with subsequent prequalification of a number of contractors. This was regarded as a reactive rather than proactive response and not in keeping with a dynamic organisation with good contacts and reputation in the tightly knit minerals industry.

Tenders, project management and contract management practices of public utilities were considered by the company to be overly complex and could create a high level of risk to contracting companies more used to negotiated, reimbursable type contracts with relatively short project duration, and clearly defined deliverables and scope. However, as most immediate project opportunities were identified within these utilities, a pragmatic and optimistic approach was adopted and personnel experienced with the particulars of these public owners were assigned to the project.

Economic drivers

Economic considerations, political changes and workload fluctuations in the company's domestic market created a need for diversification, and investing considerable time and effort to further develop the company's market segment in this relatively unfamiliar 'non core business'. This decision was further supported by specific market research, which identified a potentially large domestic environmental (natural) market over the next few years. Other significant drivers were major environmental (natural) issues associated with recent mining projects, indicating the need for in-house environmental (natural) capabilities as dictated by major owners.

Market research

The company's internal research identified existing customers that had need for environmental (natural) projects within the next few years, however few customers knew of the company's capability in this area. This indicated the need for company promotion to achieve full awareness and a higher market penetration.

Market research involved examining:
- Population growth and current facilities serving existing populations.
- Government pollution control programs, and shortlisted high priority areas and industries.
- Projected cost increases for the treatment of sewage and industrial wastewater.
- Re-use potential and possible customers for treated effluent.
- Relative activity, profitability and technology of potential competitors.
- Forecast of profits from projects.

Primary data

Interviews were conducted with key stakeholders in the industry, and in particular personnel from water and environmental (natural) authorities, council engineers, sewage inspectors and even competitors. Design consultants' capabilities were determined.

Secondary data

Third persons supplied industry background information and typical choice processes. However, the supplied information was based on a more domestic view and inward vision while ignoring recent trends in the international wastewater market.

Trade associations provided some information, however it proved to be a rather conservative commentary and not encouraging towards new and innovative design and project approaches. It also missed addressing the technological obsolescence of some of the owners' proposed designs. This finding further indicated improved chances for a company applying new, innovative and proven technologies in a historically very conservative industry.

Another approach was to identify the customers' choice process, where it was indicated that due to a change of decision making away from historically dominating 'lowest cost' approach, the final decisions were now influenced by environmental (natural) stakeholders, which included statutory authorities as well as the public. The use of 'total life cycle cost' and wide ranging environmental (natural) impact statements with community input was of paramount importance on recent water/wastewater projects.

Selected project

An upcoming sewerage augmentation project was assessed for its feasibility and an alliance with an international design consultant with highly regarded experience was put in place. This alliance was approached with the awareness that it could lead to a potential duplication of engineering disciplines, but had to be undertaken to enable prequalification of the company as head contractor.

Role of project manager

Due to the relatively small size of the company's water/wastewater group, the project manager was directly involved in all marketing and business development activities. And so a high degree of ownership was established early in the project. The project manager filled the role of bid manager, process engineer and project manager during the project initiation and tendering phase. Another key area was the close liaison with the design consultants to avoid costly duplication of engineering work and to maintain adherence to a pre-agreed project structure. The project manager's role was critical to interpret and implement these varied tasks efficiently.

6.5 PRIOR OBSOLESCENCE

The need for an asset or facility may come about through the obsolescence of some other asset. The end of an asset's life may be brought about through, for example:
• Physical obsolescence.
• Economic obsolescence.
• Functional obsolescence.
• Technological obsolescence.
• Social obsolescence.
• Legal obsolescence.
• Sale or transfer of the facility to another owner.

Some of these may be interrelated.

Physical obsolescence

Physical obsolescence implies collapse, demolishing or general physical deterioration of the facility to a point where it is no longer useable. It may be the same as the design life of the facility.

Example

Water pumping facilities

In order to determine whether new water pumping facilities are required, existing asset performance is first measured by factors such as mean time between failures (MTBF), pumping volumes and water quality. A need is identified when there is a significant gap between asset performance and customer requirements.

Different monitoring methods are used for different types of equipment. For example, breakdown monitoring only identifies a problem after a failure has occurred; planned monitoring involves checking performance at planned times; condition monitoring involves checking of the equipment regularly. The type of monitoring used depends on the criticality of failure of the piece of equipment. For example, if human health is involved, condition monitoring would be used.

Economic obsolescence

Economic obsolescence is reached when there is a cheaper alternative available.

Functional obsolescence

Functional obsolescence may be reached with a change in function of the facility, that is the purpose of the facility has changed.

Technological obsolescence

Technological obsolescence implies that superior alternative technology is available. The alternative technology may or may not be a lower cost option.

Example

Information technology

All large organisations, both commercial and government, maintain extensive and sophisticated information technology and telecommunications (IT&T) systems as support mechanisms for their operations. These systems represent a considerable investment in technology and support resources, and require extensive and continual development to maintain efficient and competitive practices. Although the systems are essential, except for IT and telecommunications companies, they are a tool and not a product or output.

Example

Computing and software

In the computer products and software industry, technology turnover is measured in months rather than years – this fact leads to earlier-generation products being physically (older computer componentry no longer being available), economically, functionally and technologically obsolete.

Social obsolescence

Social obsolescence comes about with society's changing fashions, perhaps making renovation or refurbishment necessary.

Image to owners and the public may spur a need to modernise out-of-date or old looking facilities (social obsolescence, approaching physical obsolescence); 'keeping up with the neighbours'.

Legal obsolescence

Legal obsolescence may occur with changes in government regulations and statutes, for example safety regulations and building ordinances.

Case example

Upgrade of a mine assay laboratory

There was an identified need for a new mine laboratory to be built in order to improve safety in the workplace. The safety issue involved the use of highly toxic lead oxide for the routine analysis of ores using a fire assay technique to determine their precious metals content.

Fire assaying for the determination of gold and silver involves the use of lead-based fluxes to extract gold and silver out of the mineral matrixes under high temperature conditions. All the steps involved in fire assaying have the potential of exposing employees to high levels of lead, either as airborne dust or as fumes generated during the fusion and cupellation steps. This can be hazardous if proper precautions are not taken and the building design is inadequate.

The existing fire assay laboratory was designed by a person who had little knowledge of the existing acts relating to laboratory construction, in particular those governing the use of lead in the workplace. As a result the building was poorly designed for the intended use. The laboratory had been in operation for several years.

Routine monitoring of the lead in the blood of the fire assay workers was initiated. The results established that the level of lead in the blood of the fire assay workers was twice that of the rest of the site's workforce, and twice the recommended lead levels.

Steps were taken to minimise lead exposure by wearing personal protective equipment. However this measure was only temporary, as airborne lead remained high due to dust generated within the laboratory itself and that blown in from a lead concentrate stockpile nearby.

It became obvious that other solutions were required. The justification for a new upgrade building was also supported by the possibility of increased throughput in assays without any increases in staffing.

6.6 POLITICAL INPUT

In order to please an electorate, to satisfy an electoral promise, or perhaps create immortality for the champion, a project may come about through a political directive. Special legislation may also be enacted for projects. The associated end-product may show a low benefit:cost ratio, and not be economically viable.

Under normal circumstances, without political intervention, the project may not even be given a second thought because of its low priority relative to other projects with better benefit:cost ratios.

6.7 TENDERING

For contractors, subcontractors and suppliers, the origin of a project may come about through submitting a tender (quotation, quote, proposal, offer, ...), being invited to tender, negotiating a contract, customer inquiry, or work prior to this including:
• The request for or submission of an expression of interest.
• Prequalification or pre-registration procedures.

Searching for possible avenues for submitting proposals could be partly linked to opportunity studies and market research.

Depending on the situation, the scope (of work) may be stated in a prescriptive or functional form. Overly prescriptive briefs necessarily reduce or possibly eliminate the need to look at alternatives and feasibility studies.

Case example

Redesigning welding units

A company specialising in the mitigation of wear and corrosion in industrial plants and processing plants won work through tendering. The work involved the installation of high density polyethylene liner, ancillary flashings and capping, and incorporated significant extrusion welding, much of which was on the liners.

Right from the start, it became obvious that some intelligent re-thinking of previously accepted work methods and equipment had the potential to deliver substantial benefits in both time and money and in eliminating a potentially major occupational health concern.

Existing plastic extrusion welding machines were totally lacking in any ergonomic graces. They were heavy and poorly balanced, and when extruding on floors required the operator to stoop continuously. Couple all this with the large amount of welding and long work days and there was a potentially disastrous occupational health situation. Additionally the fatigue factor for the welding operators resulted in a serious loss of productivity and quality.

The welding process was the defining process in determining the length of the contract work, and at the same time the one area where productivity increases, however small, would impact greatly on the profitability of the whole work.

With this in mind, the project of re-designing the welding units was created.

6.8 MISCELLANEOUS SOURCES

As well as the general sources mentioned above, there are a range of other sources. Some of these overlap with the general sources. Examples include:
*   The community's changing values and the community's increased awareness of the natural environment has led to a rethink of waste management facilities.
*   Immigration brings with it a need for specialist programs and infrastructure.
*   Globalisation changes the way people do business, and leads to a changed distribution of assets around the globe.
*   Religious needs create their own particular requirements.
*   Attempts at disaster mitigation promote specialist activities.
*   With competition, and to attract customers, it may be necessary to update assets, perhaps regularly; for example, licensed clubs and gaming machines, competing with hotels.
*   Equipment might need renewing and updating in order to keep it in a safe working condition.
*   Demand from users may dictate a need for an increased capacity, for example a sporting grandstand or a library.
*   Sporting facilities may have to be upgraded to be suitable for a higher competitive standard, or to incorporate more high-technology equipment.

Case example

Gaming machines

In a licensed club, gaming machines accounted for a significant proportion of the club's profits.

The purchase of more gaming machines was spurred on by a number of factors:
*   The increasing size of the club and member numbers resulting from the provision of new facilities such as a gymnasium, a pool, function rooms and a large bistro, which themselves derived from market research.
*   The potential to increase profits.
*   A government policy change allowing hotels to have access to gaming machine facilities, affecting the profitability of licensed clubs. The acquisition of more machines flowed from the club's risk management procedures of recognising potential losses.

The cost of new gaming machines is high. The percentage collection from patron input is unknown. The club acknowledges that gambling causes social and economic problems to the gamblers and their families, and provides services for patrons when it becomes a significant problem. The purchase of new gaming machines requires approval by appropriate authorities, and is helped by the club's program for the more affected gamblers.

Case example

Call centre

The purpose of a call centre is to provide a central and consistent point of contact for external telephone callers to an organisation. A call centre gives customers, potential customers, and any other interested body a point of contact to which any query can be made and a response received, without having to endure being transferred multiple times for no result.

The decision to evaluate having a call centre was arrived at for several reasons, based on anecdotal evidence:

- There was a potential loss of customers due to a lack of service. Many first time callers were not getting through to the personnel required and therefore the opportunity of gaining their business, both now and in the future was being jeopardised.
- Existing customers were not receiving the level of service they required. By not being put in contact with the correct personnel, their needs were not adequately being met. This jeopardised future business from these customers.
- There was the intangible bad will that comes as a result of someone perceiving to have received bad service.
- High paid sales staff were being diverted from their primary task of securing revenue for the organisation. This was a result of both existing and potentially new customers calling sales representatives directly, either because they had been given a particular individual's name or because the switchboard personnel transferred them to the individual based on who they believed would best be able to help. As a result, the sales staff were having to spend time locating the required personnel and/or information for this person instead of pursuing new sales opportunities. Although sometimes these calls could result in new opportunities, a majority of the time they were service related and could have been better handled by customer service staff rather than direct sales staff.

The anecdotal information was taken as being a fair indicator of a process that could be improved. Conducting a full internal cost justification would have been difficult and would have resulted in unreliable data in attempting to provide accurate financial analysis on issues such as new or repeat business lost due to a perceived low level of service.

The decision to introduce a call centre was based on a reactive response ('gut feel') to a perceived shortcoming in customer service procedures. However it was also believed that a call centre was necessary just in order to remain on par with the organisation's competitors.

The call centre would be staffed with personnel trained and equipped to handle the first level of support required to adequately service the needs of the caller. From this it was envisaged that the callers would be satisfied that:

- They had received immediate attention and were not continually bounced around the telephone system.

- Their real needs had been established quickly; this had a twofold effect in increasing customer service while decreasing the cost of the delivery of that service.
- The staff handling their call were knowledgeable in the subject matter.
- The staff handling their call had rapid access to information relevant to them.

If satisfaction could be realised in these areas, then it was believed that this would have a positive impact on the organisation's revenue from both new and additional sales, and the continued use of support and consulting services.

Case example

Excavation of a mine water storage facility and construction of a retention bund

A mine had exhausted an orebody. The mine then decided to deposit tailings in the mined-out pit, rather than raise an existing tailings dam wall. This necessitated the excavation of a water storage facility within the boundary of an adjacent orebody.

The major factors which governed operations at the mine were:
- The region was affected by a monsoonal climate with distinct wet and dry seasons. Due to this climatic regime, water management dealt primarily with storage during the wet season and disposal throughout the dry season.
- Considerable legislation influenced the operations at the mine, particularly those related to water management, treatment or disposal.

This project was a subproject of the long-term development of the adjacent orebody. The project included:
- The excavation of a water storage facility to provide surge storage in the wet season.
- The construction of a retention bund to protect the adjacent orebody from inundation by a nearby creek. The bund would also become the main site access road prior to full scale mining to alleviate traffic problems associated with a haul road crossing a general access road. The excavation of the water storage facility would provide material suitable for the retention bund construction.

In the past, the mined-out pit was used for surge storage during the wet season, with mining continuing unaffected on the upper benches. At the end of the wet, the pit would be dewatered to a retention pond as volume became available through mill consumption and disposal of water via approved methods (primarily spray irrigation). At the commencement of the tailings deposition in the mined-out pit, the pit would cease to be available for surge storage of pondwater because the two water types – process water (tailings) and pondwater – needed to be segregated due to a significant difference in water quality. Direct release to the creek was to be avoided.

EXERCISES

1. Review the strategic plan of your organisation. What projects have resulted from, or will result from, this plan?

2. A person is contemplating building campervans for hire to holidaymakers. Before embarking on this project, what information would you need to know about:
- Competition.
- Market.
- Product differentiation.
- Profitability.
- Seasonal effects.
- User characteristics, patterns, requirements.

By what method would you collect data on the above issues?
   What information would you need to know before proceeding with the project?
   How does the market research influence the scope of the project?

3. A person is contemplating building a child care centre. What information should go into the opportunity study.

4. The three fundamental business questions that organisations may ask themselves are:
- Where are we now? (Growing awareness of changing customer needs, worsening competitive situation, ...)
- Where are we going? (Based on the organisation's vision, shareholder requirements, identification of industry trends, choice of market, ...)
- How do we get there? (Evaluation of alternative solutions, marketing strategy for the final product, ...)

In what way might answers to such questions lead to new projects?

5. Give examples, of which you are aware, of each of these forms of obsolescence:
- Physical obsolescence.
- Economic obsolescence.
- Functional obsolescence.
- Technological obsolescence
- Social obsolescence.
- Legal obsolescence.

6. Identify projects in your neighbourhood that have come about through a direct result of political pressure.
   Identify examples of special legislation that applies to specific projects.

7. For work that comes about through tendering, does this reduce the amount of work in the initiation phase, or does it only change the emphasis on the different activities in the initiation phase? Discuss.

8. Suggest ways (other than given above) that projects may come about, based on your experience.

CHAPTER 7

Objectives and Constraints

7.1 PROJECTS AND PROJECT OBJECTIVES

Project v end-product

There will be an identified need or want for some product, facility, asset, service etc. This end-product is achieved through a project.

This distinction sometimes causes people confusion and many people are not aware of the distinction, or of the need to make a distinction. For example, people sometimes refer to a building as a project. It is not. The processes that go together to materialise the building are the project. (It is acknowledged that the definition of a project is sufficiently flexible to include the operation and maintenance phase of a product within what is called the project. However, this is not the issue here.)

Project objective v end-product objective

The materialisation of the end-product can be performed, possibly, in an infinite number of ways. In systems studies terms, the problem is an inverse problem, and hence there is no unique solution (Chapter 16). Design problems and planning problems are two well known other inverse problems.

The criterion by which the preferred materialisation of the end-product is selected is the project objective(s). Scope and resource considerations on projects follow.

In a similar idea, there are possibly an infinite number of versions of end-products. The criterion by which the preferred end-product is selected is the end-product objective(s). Form, function, finishes etc of the end-product follow. The selection of the preferred end-product form is a design problem.

That is, on any project there are two types of objectives:
- End-product objective(s).
- Project objective(s).

See Figure 7.1.

Commonly, project objectives say something about project cost, project time and deviation from specification, but other objectives are possible. And these may apply throughout the project (for example, minimum deviation from specification), or at the final (terminal) point of the project (for example, minimum completion time). That is,

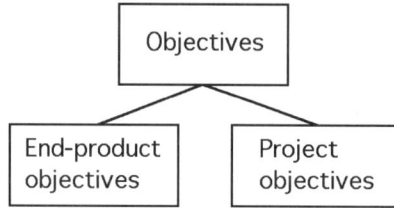

Figure 7.1 End-product and project objectives.

general project objectives will contain a component over the time domain of the project and a component at the final (terminal) point. See Chapter 17.

End-product and project objectives may derive from higher level values within an organisation, for example originating from a corporate plan. Such values may also reflect political, marketing, environmental (natural), ... concerns.

End-product objectives and project objectives (and constraints) may relate for example to:

- Money (end-product – sales, benefit:cost ratio (BCR), net present value (NPV), ...; project – cost, budget, ...).
- Time (end-product – lifetime, ...; project – timescale, ...).
- Quality issues.
- Community acceptance.
- Environmental (natural) effects.
- Safety.
- Risks.
- Public utility impact.
- Extreme event impact (floods, cyclones, ...).
- Social impacts.
- Geotechnical considerations.

These are expressed in terms of end-product matters or project matters, as the case may be.

Interchangeability or confusion?

Confusion can arise when people start expressing project objectives in terms of the end-product requirements or end-product objectives.

For example, if the end-product is a road of low maintenance and capable of carrying heavy vehicles, then the project objectives are not:

- (minimum) Maintenance costs.
- (maximum) Load carrying capacity.

or similar. The project objectives should relate to how the work of designing and building the road is to be carried out.

Further examples of objectives which are not project objectives but rather end-product objectives are:
- (minimum) User operating costs and time.
- (maximum) User safety.

End-product objectives are important and influence the way project work is carried out and can influence project objectives, but they are a separate issue to project objectives.

An example dilemma people can get themselves into, if project objectives are confused with end-product objectives, is when, say, the project objectives are written in terms of some maintenance requirements of the end-product. The question then becomes, when does the project end – after maintenance? – in which case the project never conceivably ends until the product has reached the end of its useful life. It is possible to define a project as ending at the expiry of the end-product's useful life (the definition of project is sufficiently flexible to allow this), but this is not conventional. Usually a project is defined to end as maintenance/operation starts.

Lay usage of 'objective'

Usage 1
The problem is believed to occur because there is a distinction between the strict definition (as given here) needed to perform problem solving in a systematic fashion, and the loose lay person or dictionary definition of the term objective. The latter is given as 'the object of one's endeavours'.

If the object of your endeavours is 'to build a road', this doesn't help either the project management or determine the final road. An infinite number of project approaches and end-products can fit this description. The objective has to be made more definite, that is, given a tighter prescription.

'To build a road' or 'to develop and supply equipment', for example, are not objectives, but more akin to scope statements; 'scope' and 'objectives' are becoming interchangeable terms through loose lay expression. A number of writers wrongly say that objectives are a 'statement of work'.

Usage 2
There is also a lay person's usage of the term objective as being a target or a destination, that is the point where you wish to finish up. For example if you are travelling from A to B then point B is your objective. Some management texts talk in terms of a 'goal' – 'what is wanted from the project'; 'something that gives direction to the project'; 'a word picture of the end result'; 'what must be met for the project to be considered successful'.

In this sense, an objective is akin to a deliverable (end-product plus state – See Chapter 16 – final (terminal) conditions for the project), key result area (KRA), key performance indicator (KPI) or similar. [Apologies for using these trendy terms.] A number of writers wrongly say that objectives are deliverables.

Some pseudo rigour may be given to this by listing the characteristics of an objective to be, for example:
- Clear.
- Concise.

- Achievable.
- Realistic.
- Relevant.
- Understandable.
- Agreed.
- Consistent.

Or extra pseudo rigour may be given by making the characteristics fit a catchy acronym, for example:
• Specific.
• Measurable.
• Agreed.
• Reachable.
• Timely (or Time-Bounded).

If it is a desire to undertake a project for cost C, in time T, according to a specification S, the better way to write the project objectives is in terms of minimising the difference between actual cost and C,

$$\min |c_{actual} - C|$$

minimising the difference between actual time and T,

$$\min |t_{actual} - T|$$

and minimising the difference between actual work standard and specified levels S,

$$\min |s_{actual} - S|$$

Alternatively, if it is a desire to undertake a project for a cost less than C, in a time less than T, and to a standard better than S, then these are constraints and not objectives. Such constraints can live simultaneously with the above stated preferred forms of objectives; they may even be redundant.

Usage 3
Many lay people further make a distinction between an aim and an objective. There an aim is regarded as a statement of general intent. For example the aim of this book is:
• *To develop an awareness and understading of project management.*

In contrast an objective states requirements in precise terms, for example:
• *On having completed this book, readers will be able to correctly undertake 50 different project management, practices.*

It is in this sense that the term objective is used by many writers, and not in the sense needed for project management purposes.

It appears that the word objective is used by most people in the sense of Alice in Wonderland:

> 'When I use a word,' Humpty Dumpty said in a rather scornful tone, 'it means just what I choose it to mean, neither more nor less'.
> 'The question is', said Alice, 'whether you can make words mean so many different things'.
>
> ('Through the Looking Glass', Ch. 6, Lewis Carroll)

Case example

Selection of a faxing system

Senior management wanted to leverage their investment in personal computers (PCs) and a local area network, to provide a faxing capability for the office personnel.

Cost was not a factor. All of the proposed solutions were similar in cost. It was felt that the savings in paper, savings in time hassling with faxes, and the better security considerations would pay for the PC fax product within six months. The emphasis was on the quality and consistency of the selected PC fax product. It would have to be robust enough to handle up to 500 users. Once the company personnel started using the product, they would not tolerate its unavailability due to unreliable software or hardware.

Six products were reviewed through the literature. Four products were evaluated hands on.

The 'objectives'

The overall 'objective' was stated as:
• *'To provide a PC-based faxing solution.'*

Specific 'objectives' were stated as:
• *'Include outbound fax addresses in address lists.*
• *Fax from any PC application that can print.*
• *Deliver an inbound fax for individual users without human intervention.*
• *Good technical support. Once the PC fax software was deployed, many people in the company would become dependent upon it. The company wanted a fax solution that the local supplier could maintain efficiently and quickly. The company was buying a solution, not just some fax software and a few modems.*
• *Good system administration tools. The help desk people must be able to rectify lost faxes, trouble shoot a user's problem etc.*
• *The PC fax software should be very easy to use and not require much training.'*

Confusion

Clearly the company has confused objectives and constraints and scope. The specific 'objectives' were to be used to make the final selection of the product.

There is nothing in these 'objectives' which helps discriminate between alternative products, assuming all products meet the specified constraints. What may have been intended was a fax system with the best system administration tools, the best technical support etc. On pointing this out to the company, the 'objective' was revised to the following:

- *'The objective is to deliver fax capabilities to the desktop via a user's PC. The users must be able to directly send and receive faxes from their PC without the users needing to use a fax machine. The PC fax product must be robust. It must handle up to 500 Users. It must be easy to use.'*

The company's comment was: *'The objective [now] is a simple statement detailing the vision of the selection process. It gives a clear and concise mission statement. If anyone should ask what the selection process is all about, one can simply quote the objective. It includes a commitment to performance that the selected product must be robust (a measure of 'industrial strength') and be able to cater for up to 500 users. The users and the person, who is making the selection, can objectively agree on what is being delivered.'*

Clearly this is no better than the first 'objectives' proposed. There is still nothing here to help discriminate between alternative products. The content of the 'objective' is only in terms of requirements that have to be satisfied by candidate products. The company is just using the word 'objective' because everyone else in management uses the word objective. The company has no idea what an objective is. A similar comment could be made on the company's use of the terms 'vision' and 'mission'. It appears that the company, like many people also thrown into management, use trendy terms to disguise ignorance.

Case example

In-house software development

The case study project is an in-house software development project – delivering an (owner) problem tracking system. This project had appointed a project manager and owner's business representatives early, in order to align technology and business interests.

The company listed the following project 'objectives':
- *'Estimate and control cost.*
- *Estimate and control delivery time.*
- *Provide a quality solution.*
- *Replacement of an existing system exhibiting poor reliability and low functionality.*

- *Provide a solid system to support the business needs.*
- *Utilise newly proposed technology and methods as directed by the organisations' strategic technology planning group.*
- *User community acceptance.*
- *Acceptable cost/benefit.*
- *Provide some business process reengineering (BPR) consultancy.'*

The company's reasoning behind this was as follows.

'*The initial solution proposed, had a no added functionality (NAF) stigma attached. This is where systems are reproduced functionally verbatim to replace an older system. As a differentiator, it was decided that some BPR consulting would be provided to ensure higher system availability, and to streamline and automate some business processes.*

The company was beginning to place focus on quality standards and methodologies. This project would follow an internal project development methodology that was also supported by a computer aided software engineering (CASE) tool.

Cost control was a major objective. Any change in the cost would affect the economic feasibility. The business also indicated that there was an urgent need for this system, as failures in the existing system had caused the company embarrassment, loss of trust and professionalism when dealing with customers. Thus follows, a priority objective to meet delivery times.

It is important to note the primary relationships involved that affected some objectives identified. Project teams were responsible for providing business solutions. The strategic technology planning group was not associated with any particular business line; its role was to ensure that the technology could support the project teams and hence the business solutions. The group imposed on this project a new technical direction which involved a significant skill-set and paradigm shift for staff as well as the company as a whole.'

The company recognised that the 'objectives' were poorly defined, and that the project would be adversely affected by the multiplicity of 'objectives'.

The 'objectives' are clearly not all project objectives, and those that are could be better expressed. The question then follows of what is the influence on the project direction of not formulating project objectives properly, and hence what is the point of listing such items at all.

7.2 ALTERNATIVE NAMES FOR OBJECTIVES

As needed to perform systematic problem solving/systems synthesis, alternative names include:
- (Optimality) criterion.
- Performance index.

- Performance measure.
- Merit function.
- Pay off function.
- Figure of merit.
- Aim.
- Goal.
- Cost function.
- Design index.
- Target function.

These generally occur outside the management literature. They are common in the optimisation, optimal design and optimal control literature. (Carmichael, 1981)

Their usage (together with terms like mission, key result area (KRA), key performance indicator (KPI), vision, ...) within the management literature is for purposes of impressing the reader as jargon rather than to help the management process. Their usage in the management literature is generally not consistent with any usage for systematic problem solving/systems synthesis.

7.3 CHARACTERISTICS OF OBJECTIVES

Wherever possible, objectives are stated in *measurable* terms in order that a comparison of alternatives can be made quantitatively.

The other important characteristic of an objective is that it be *capable of being extremised* (that is, minimised or maximised).

Objectives might be classified as:
- Owner generated.
- Project generated.
- Externally generated.

and

- Essential.
- Desirable.

or given different weightings or priorities.

Example

Commonly objectives, either for an end-product, or its associated project, are seen to be stated (wrongly) in the following forms:
- To satisfy a political undertaking.
- To demonstrate a government commitment to development.

- To reduce soil erosion.
- To ensure availability and continuity of funds.
- To keep the community informed of progress and expected outcomes.
- To reduce user operating costs and times.
- To increase user safety.
- To reduce maintenance costs.
- To increase load carrying capacity.

It is argued that in such forms they are stated incorrectly, because of the non-uniqueness of the solution that results where the objective is not capable of being extremised.

The above examples might be more correctly termed aims or goals which might apply at a level higher than the project or end-product. They lack definiteness. Some may even be interpreted as constraints.

7.4 ORGANISATION HIERARCHY AND OBJECTIVES

In terms of a large public-sector or private-sector organisation, the organisation's hierarchy might take the form of Figure 7.2.

The organisation has broad aims which might be called (loosely) goals, key result areas or similar. These aims are passed down the hierarchy in Figure 7.2 becoming more definite as the level drops. The term 'tree out' may be used when expanding the organisation goals into individual goals. This is a top-down approach.

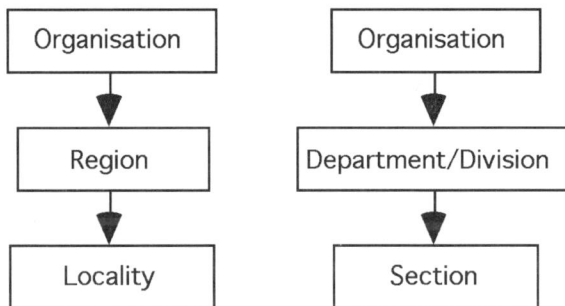

Figure 7.2 Organisational hierarchies.

7.5 MULTIPLE OBJECTIVES

For any situation there may only be a single objective or there may be many objectives (multi-objective). Where there are multiple objectives, they may be *non-commensurate* and *conflicting*. One or more of the objectives may be more dominant than the others.

The solution of such multi-objective problems generally involves some subjectivity while the solution of the single objective problem does not.

Where multiple objectives exist, there are three broad approaches which can be adopted:

- Assemble all objectives into a single objective using, for example, weighting functions.
- Only consider one of the objectives as the true objective and treat the remaining objectives as constraints.
- Develop an independent solution for each objective and trade off the multiple solutions.

Each approach involves some subjectivity.

Weighted evaluation using, for example, an evaluation matrix is a common approach. The choice of weights needs to be justified.

The reinterpretation of some objectives as constraints, that are not to be exceeded or gone below, is attractive to many people's way of thinking.

Example

Assume a facility is required to be both minimum cost and to give maximum production. Cost has units of money, production has units of output/time.

A combined objective may look like:

(min) w_1 × cost + w_2 × production

where w_1 and w_2 are weighting functions subjectively chosen with full knowledge of the different units of measurement involved, and reflecting the relative importance of each objective, and whether minimisation or maximisation of objectives is required.

A single objective together with a constraint may look like:

(min) cost
(subject to) production ≥ specified amount
or
(max) production
(subject to) cost ≤ specified amount

7.6 CONSTRAINTS

All projects have genuine constraints such as funding, environmental (natural), and political. Constraints limit the range of decisions possible; they restrict the options that are possible.

As with objectives, many people confuse a project constraint with an end-product constraint. End-product constraints may influence project constraints.

Constraints might be classified as:
* Owner generated.
* Project generated.
* Externally generated.

and

* Essential.
* Desirable.

or given different weightings or priorities.

Consider a comparison between constraints and objectives.

Example project objectives

* (minimum) Impact of project activities on the natural environment.
* (minimum) Project time.
* (maximum) Utilisation of resources.

Example project constraints

* A certain specified level of interaction with the community.
* Availability of funds up to (money).
* Complete the work by (date).
* Complete the work for less than (money).
* Do the work with less than (resources).
* Hazards and local factors.
* Statutory requirements.
* Allowable noise levels.
* Site space restrictions.
* A lower limit on the standard of work.

Example

Safety constraints

The mining industry identifies safety as being one of high priority. Many initiatives are undertaken to promote safety awareness and ensure the safest possible workplace. All stakeholders become involved.

　　All construction, operation and maintenance are undertaken constrained by safety issues.

Constraints, of course, can be converted to objectives (and vice versa). For example, if it is desired to complete a project in less than 100 days, this constraint can be turned into an objective that minimises the difference between actual time and 100 days, that is

min | t – 100 |

Similarly, other constraints, for example related to expenditure limits, can be turned into objectives.

Example

Project cost limit

A project cost limit is an upper limiting cost constraint set by the owner.

　　There may also be project duration limits.

　　These limits may be in excess of what the project team has to work within, and may represent limits acceptable to the level of management above the project level.

7.7 PERFORMANCE MEASURES

Performance measures are gauges by which the success or otherwise of a project is measured. A performance measure is in the lay usage sense of the word objective.

　　If the project is regarded as a system, then the system's output is the performance, and is measurable (Fig. 7.3).

　　Generally a project has a planned-for performance (in terms of cost, time, quality, productivity, ...), and actual performance is compared with planned performance. A project's success is then measured in terms of a *variance*, that is, the difference between actual and planned.

　　Performance measures tend to be developed later in projects; objectives are developed very early. A performance measure could be expected to follow planning and develop-

Input/
control

Output/
performance

Project

Monitor

Feedback

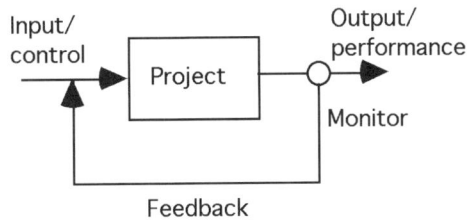

Figure 7.3 Project as a system.

ment of baselines (planned levels) for cost and time, and quality plans and productivity expectations.

A performance measure might be termed a key performance indicator (KPI) or similar, but all of these types of terms are used very loosely by most people. The terms are used frequently by project personnel because they sound good and impress, but lack precision.

Case example

Intelligent building

The project centred on a subcontract involving the supply of hardware and software for building management, security and photo-ID for a prestige building.

The project was to develop and supply equipment and software as specified in the (sub)contract negotiated by the subcontractor. This was to be done for a price that enabled the subcontractor to win the tender, while still returning a profit for the company. In addition, there were returns resulting from the high public profile of the building. The returns related to public and industry exposure were to have highest priority.

Objectives and constraints

The objectives and constraints as stated by the company follow. They contain some information on scope, would require reworking to distinguish project issues from end-product issues, and would also require reworking to get them into useable objective and constraint form.

- *'A top-quality system with maximum public exposure of the completed system.*
- *A photo-ID system that fully meets the end-user requirements and is coordinated with the other components of the project – that is, building management system and security system.*
- *A system delivered at or below the quoted price.*
- *Requirement to avoid negative publicity.*

- *Requirement to appear as fully professional and competent in all respects at all times.*
- *Significant restrictions on negotiating with the end-user to modify requirements.*
- *Requirement to conform to installation schedule dictated by other subcontract work at the same site.'*

Origin of objectives and constraints

Some of these objectives were dictated by the contractor's strategic plan, in which this building project had been identified as a critical component. Some of the constraints arose from the size, visibility and complexity of the project.

The listed objectives and constraints were largely taken for granted and accepted as a necessary result of the strategic significance of this building work.

Case example

Development of an industrial estate

The project referred to in this case study was the development of an industrial estate. In this case, the owner/developer had specified the end-product as a number of light industrial lots, completed with local authority approval, and legally gazetted to allow contracts of sale to proceed. The project was the process which brought about this end-product.

Although not particularly complex, the project was fast-tracked, and the lead time given to the project manager prior to commencement was short. The work was mobilised earlier than scheduled in the developer's master program. This decision was brought about by unexpected demand for the particular product, in this case being completed industrial lots, approved and ready for sale.

The project management services were provided by a civil design consultant. His brief was to also provide planning and design services from conception through to approvals, provide contract administration duties and liaise with and coordinate sub-consultants. The contractor and sub-consultants were directly contracted to the owner. The contractor was accepted by the owner on the basis of its performances on numerous previous projects and agreed rates, with formalisation of the contract (involving a lump sum amount) to follow.

The scope (of work) involved the filling of four adjacent areas to a nominal fill level and completing the services (roads, stormwater pipes, sewerage etc.) to one of the areas. The bulk fill work had previously been approved by the local council, but at the time of commencement, no approvals for the construction of services had been obtained.

Site geotechnical reports had previously been received and these highlighted the presence of marine mud at depth in isolated zones. Preload, to assure stability for foundations of future buildings, was recommended at those zones with underlying marine mud.

In summary, the project commenced under the following conditions:
- Part of the work approved by the local authority.
- Design 80% complete with the balance of local authority approval to follow.
- 'Ceiling' cost based on preliminary estimates by the consultant and adopted by the owner as the project amount.
- Contractor given possession of site; contract price – to be finalised; documentation to be completed.
- Date of completion specified by owner.
- Preload removal dependent on rate of settlement and recommendations by geotechnical consultant.

Project objectives

The project objectives at the time of receiving the direction to proceed, differed to the anticipated objectives corresponding to the original program. The fast-track situation involved incomplete design and incomplete earthworks.

Objectives created by the owner – the owner was interested in the timely sale of the lots and this created the objectives of time of completion and budget:
- (minimum) Duration of the project.
- (minimum) Cost.

Other objectives were not a direct concern of the owner but were objectives of concern to the project manager:
- (minimum) Time to complete designs.
- (minimum) Set up time for tender documents after receiving approvals.
- (minimum) Duration of preload.

Project constraints

- Work to be done as directed.
- Quality; maintenance of accepted design and construction standards (marketing requirements) and within approval criteria of the local authority.
- Maintenance of scope in accordance with that previously submitted to the owner.
- Documents appropriate to local authority approvals and conditions.
- Act on geotechnical recommendations following investigations.
- Budget (maximum price).
- Completion date (specified by the owner).

End-product constraints

- Industrial lots, suitably finished to enhance marketing.
- No drainage discharge problems.

- No detrimental effects on existing neighbours.
- An acceptable level of impact on the natural environment.
- Zero defects or omissions during the defects liability period, particularly for the purpose of final acceptance by the local authority.
- Long-term durability so as to maintain the good name and integrity of the developer; this would enhance future marketing opportunities and continue goodwill in dealing with the local authority.

Case example

Communications outsourcing

All large organisations, both commercial and government, maintain extensive and sophisticated information technology and telecommunications (IT&T) systems as support mechanisms for their operations. These systems represent a considerable investment in technology and support resources, and require extensive and continual development to maintain efficient and competitive practices. Although the systems are essential, except for IT and telecommunications companies, they are a tool and not a product or output.

Many of these organisations consider transferring their systems and the responsibility for support and development to specialist, external service providers.

There are several factors that have contributed to considering outsourcing as a viable alternative to in-house investment in and support of IT&T services:

- Technology. Increasing complexity and sophistication of technology requires greater investment in specialist support to manage and develop systems and services.
- Competitive pressures have encouraged organisations to focus on core business activities, and to avoid diverting time and resources into what can be considered as side issues.
- Costs. Computing and communications are integral to operations and so have attracted an increasing proportion of capital and operational budgets. Options for managing the costs are being increasingly considered.
- Alternatives. The number of companies offering outsourcing services has increased and the industry has matured to a point where there is growing confidence in the market's ability to support IT&T outsourcing requirements.

This case study looks at projects to provide the services from the contractor's viewpoint.

IT&T outsourcing projects carried out by the contractor are medium- to long-term operations, often contracted for several years and sometimes extending to ten years. The projects are also multi-faceted, combining elements of design, construction, implementation and ongoing management. They can be comprised of computing and/or communications, customer equipment, public and private networks, service management including diagnostics and repair, billing, inquiry and help-desk functions. The scope of the project is also subject to change during its life cycle, particularly for longer-term projects due to changes in technology, regulation, competition etc.

The objectives over longer terms can become dated or even irrelevant should the focus of any of the stakeholders change. Large projects frequently require a long-term commitment to make them commercially viable. It is critical that the longer-term objectives are carefully considered and understood.

The end-product is an externally provided and on-going supported service with the objectives of:

- (minimum) User costs.
- (maximum) User service.
- (maximum) User access to higher levels of technology and expertise.

These objectives are determined by the owner – the organisation that is outsourcing its systems and services.

End-product (service) constraints:

- Cost per unit (equipment, network traffic etc).
- Service availability (for example, mean time between failures, network down times, fault response times) (quality).
- Timeliness.
- Agreed standards.

The contractor's project objectives are:

- (maximum) Additional business.
- (maximum) Profit.

The contractor's project constraints are:

- Retention of owner loyalty.
- Bid value.
- Geographic – owner locations that test a service provider's coverage. This is often overcome through partnering arrangements with third party suppliers.
- Technical – support for some components of the technology outside of the expertise of the service provider. Subcontracting addresses this issue.
- Access to the appropriate specialist expertise and support manpower.
- Provision of appropriate technology.

Retention/development of business and maintenance of margins can be conflicting if the service is considered in isolation. Those issues may be evaluated in the context of the owner's total business with the contractor, and not just in terms of a single project.

Case example

Plane vibration testing and reduction

A company was approached by its aerospace subsidiary to help it troubleshoot a problem. The background information was that the aerospace subsidiary had numerous complaints from pilots that abnormally high vibration and noise was being developed in a particular aircraft.

A detailed discussion with the pilots was held to determine the history of the problem plane and the changes that had gone into the plane to reduce the vibration and noise. Information given was that the aerospace subsidiary had changed and replaced the plane engine from that in a similar plane which had no vibration and noise problems. The problem plane was flown again by different pilots and the same complaints were made. Non-destructive testing of the structural integrity revealed no defects. Physical dimensional checks on the problem plane found dimensions to be within the plane's geometrical tolerances established by the manufacturer.

With this limited but useful information, the company was asked to find out what was the cause for the high vibration and noise levels on this problem plane, and what could be done to reduce the levels in a cost-effective way.

Subsequent meetings were held. The owner wanted to have a general survey of the plane done when the engine was throttled on the ground. The owner wanted preliminary ground test data before agreeing to an actual flight test.

The end-product

The objectives and constraints for the end-product might be described in terms of the following.

End-product objective:
• (maximum) Plane safety.

End-product constraints:
• Human perception levels for noise and vibration below complaint levels.
• Limits on structural vibration levels.

However, all was not smooth on the project. The objective and constraints were left indefinite. Several changes in management decisions caused the focus to shift. This made the project difficult to progress. Human perception levels of vibration and structural vibration were ranked equally important. This created difficulties in prioritising resources to carry out the test. Instrumentation for structural vibration measurement is different to that for human vibration measurement.

The objective and constraints were not quantified. The amount of reduction to be achieved in vibration and noise levels before the project could be considered successful was not indicated. What should be used for the evaluation of human vibration was not stated. What might be considered acceptable to the project team may not be agreeable to the owner.

The project

The objectives and constraints for the project might be described in terms of the following.

Project objective:
• (minimum) Cost.

Project constraints:
• Budget.
• Duration.
• Availability of planes.
• Logistical issues.
• Availability of air-space clearance.

The project constraints were not defined clearly. The overall budget allocated for this investigation project was not indicated. When a 16-channel data recorder was proposed for use in order to increase the number of recording points, the additional rental charges were found to be high and the proposal did not proceed.

There were numerous engineering methods that could be applied to reduce the vibration and noise levels. These could range from placing noise/vibration damping material, tightening loose bolts and nuts, lubricating moving joints, strengthening a structural member, to a full scale experimental modal testing and analysis. The problem could be solved quickly and cheaply or it could require long manhours to perform the modal testing on a full scale plane. The actual method used (and hence the cost) would not be known until after the testing.

The time allocated to complete the investigation was also not defined. The owner and the aerospace subsidiary had internal disagreements and a conflict of interest with regards to the problem plane. They contemplated whether the investigation should be carried out at all, considering the fact that the problem plane was already close to its recommended maximum flying hours and it would have to be scrapped when this happened. The project team wanted to carry out the investigation immediately. However it was faced with numerous constraints such as the non-availability of the planes, the logistics of preparing the plane for testing, and delays in seeking air-space clearance.

The owner's changing management resulted in long times to get project approvals. This resulted in the project not having a clear starting point and hence the target date for project completion was not known.

Vibration measurement is basically a survey technique used to map out the vibration severity in the structure of a plane. As no definite source of vibration could be identified by the pilots, a number of sensors were placed at the most likely positions where vibration is transmitted and magnified. The amount of useful data that could be captured was therefore constrained to where the sensors were placed. This limited the investigation because there was a possibility that there were parts of the plane having severe vibration which would not be detected at all by the instrumentation because the sensors were not placed there. Hence there was the possibility that high vibration levels would not be picked up and hence an effective vibration/noise reduction program could not be achieved.

Case example

Site remediation and recreational facilities

This case study concerns a waste project. In broad terms, the project involved the site remediation of a former tip site with the view to establishing outdoor recreational facilities and waste management facilities. Previously the site earned a marginal revenue through commercial activities on a small portion of the site.

   Here, it is important to draw the distinction between the project (site remediation and facilities development) and the end-product (outdoor recreation facilities and waste management facilities) as objectives and constraints exist for both the project and end-products.

Project objectives included:
- (maximising) Workers' safety during the project.
- Cost (minimisation).
- Time (minimisation).

End-product objectives included:
- (minimisation) Of the environmental (natural) risks associated with the site.
- (maximum) Community acceptance and social benefits of waste management operations on the site.
- (maximum) Benefits to the local community and general public through the provision of recreational facilities and value creation for the local community and council (social and financial) from non-waste management activities conducted on the site.
- (maximum) Area of structurally sound land zoned industrial.

End-product constraints included:
- A commercially acceptable return to the owner through its site activities.
- A building height limit.
- The site's prior classification as unhealthy building land.
- The generation of contaminated leachate from old tip fill and an adjacent water body.
- A road reserve which covered the entire site and its rationalisation.

- Meeting or exceeding guidelines for remediated land.
- The geotechnical conditions of the site and the nature of the contaminated fill material.

Project constraints included:
- Government approval of the site remediation action plan, and recycling activities on the site.
- The council and local community's concurrence on waste management activities.
- The effects on the owner's other waste management operations.
- The effects on local catchments.
- The approval of government waste authorities.
- The transfer of land between authorities.
- Provision of an adequate temporary space.

Failure to satisfy any of these constraints would see the abandoning of the project. Local council expressed its desire to commence the initial phase of the community consultation prior to the land transfer and the rationalisation of the road easement; failure to achieve either of these would mean the death of the project. On the other hand, early community consultation would help to win community acceptance of the project and end-product. It was thus decided that negotiating the land transfer and road reserve rationalisation would be followed by the community consultation and finally the site characterisation (geotechnical, contamination, groundwater and gas investigation), on the basis of the initiation budget.

Gaining community acceptance and involvement, whilst not quantifiable, was vital to the success of the venture. Extensive and comprehensive community consultation was planned and budgeted. The local community was suspicious of major developments and further development of the area may have generated considerable hostility.

There were also a number factors which, whilst not being constraints, proved to have a large influence on the project and end-product. These factors included political support in principle for the establishment of a 'world-class waste recycling and disposal facility' on the site, the rumour of an imminent election, and the government's move to focus on new waste treatment technologies such as waste-to-energy and composting.

As the project progressed, a number of additional constraints and objectives came to light as a result of the increase in the number of stakeholders and their involvement. Stakeholders included a number of local, state and federal government bodies, a road authority, water and waste authorities, and local community groups. With the increase in the number of stakeholders, scope creep occurred. The objectives and constraints of each stakeholder differed.

Although the objectives seem relatively straightforward, the number of constraints on both the project and end-product made the project difficult. Furthermore, whilst the project constraints were largely engineering in nature, the end-product constraints were more socially-oriented in nature and involved a strong consultation component. This followed from the owner's function as a service provider to the community:

'*We are committed to involving the community in planning and implementing decisions which affect them. We are committed to providing and maintaining healthy and safe conditions for all employees in their workplace and for customers and visitors to our facilities.*'

The objectives were also in line with the owner's corporate aims to provide quality waste management services and solutions with a commercial return to shareholders.

Case example

Factory building extension

A factory was located on half of the owned land. The factory consisted of two bays, office building and parking. The factory produced medium to heavy-medium specialised engineering products, and machined components which were assembled into modules. These were fully assembled and commissioned in the factory. Therefore, a large floor area was required because the machinery was normally large.

The manufacturing processes comprised machining, fabrication and assembly. The machine shop totally occupied one bay and part of the second bay, along with the welding and assembly process. The sharing of the second bay prevented further expansion of the machine shop, and there was interference of the fitting and welding processes with the machining process.

The business required more room for more efficient construction and to enable more modules to be built at the one time. The extra production area would allow turnover growth and the possibility of installing new processing machinery.

The company had the vacant land to extend the factory. There was no additional capital outlay needed to purchase extra land and the vacant land was a resource which was not being used. The factory was to be designed and built by business employees, with the major expenditures being the earthworks and raw materials. The project was seen as having a long-term payback.

The project need was to end the existing space limitation, and also to enable the company to expand the number of process machines and become more efficient in manufacturing. This would lead to opportunities for larger work, which required greater assembly and manufacturing space.

End-product constraints

- Efficient work area, less crowding.
- Good head height.

- Large open areas.
- Two overhead gantry cranes.
- Electrical, air and water outlets positioned throughout the factory.
- Good natural light (power saving).
- Entrances suitable for larger loads.
- Cost.
- Environmental (natural) and planning regulations (usage, car parking, storm water detention and pollution, beautification).

End-product objectives

- (maximum) Value for money.
- (maximum) Efficiency.
- (maximum) Return on investment.
- (maximum) Increase in value of land.

Project constraints

- Cash flow. (This led to the building being constructed in two sections over a period of two years. The whole space was not needed instantaneously.)
- Community acceptance.
- Environmental (natural) and planning regulations (operation hours, construction details, waste).

Project objective

- (minimum) Cost.

Case example

Upgrade of a mine assay laboratory

A new mine laboratory was proposed to be built in order to improve safety in the workplace. The safety issue involved the use of highly toxic lead oxide for the routine analysis of ores using a fire assay technique to determine their precious metals content. The justification for a new upgrade building was also supported by the possibility of increased throughput in assays without any increase in staffing.

A design company was given the following end-product objectives and constraints for the building design.

End-product objectives

- (minimum) Lead levels in blood of fire assay workers.
- (minimum) Congestion.
- (maximum) Productivity.
- (maximum) Safety at work.
- (maximum) Improvement in general working conditions.

End-product constraints

- Conforms to the design specifications.
- Reduce airborne lead to less than a defined level.
- Compliance with standards and legislation in relation to airborne lead.
- Cost had to be within budget.
- Building to be built on current location after demolition of the existing laboratory.
- Building had to fit in the confined space available.

Case example

Redesigning welding units

A company specialising in the mitigation of wear and corrosion in industrial plants and processing plants won work through tendering. The work involved the installation of high density polyethylene liner, ancillary flashings and capping, and incorporated significant extrusion welding, much of which was on the liners.

The project required the re-designing of welding units. The new welding would be expected to:
- Eliminate potential occupational health problems.
- Increase welding consistency.
- Increase welding rates.

The following constraints were placed upon the project and end-product:
- Weld quality could not be lowered.
- Weld rates had to increase to a level that would financially justify any capital expenditure on the project.
- Any equipment purchased/ fabricated or in any way used had to be able to be replaced or repaired with minimum disruption to the on-site program.
- Welding machines were to be used in very remote locations and therefore needed to be robust.

Staff subsequently undertook some hurried studies in plastic extrusion welding, its methods, evaluation techniques and requirements and test requirements.

Weld parameters and preliminary design requirements were established. An evaluation was undertaken to determine the welding rates required to justify the best guesstimate of costs. It was very apparent that using off-the-shelf equipment would provide the best course of action with regards to costs and maintenance of the units.

Once the requirements were established, it became a relatively easy task to find acceptable equipment, alter it to the company's needs and do preliminary testing.

The final result combined the original ungainly, and yet fully functional, weld machine with a small, very robust, battery or mains powered tractor that had a precisely controlled electrical motor. The weld machine was mounted on linear bearings and its weld pressure was controlled by the addition of a counter weight. The speed was controlled by means of a hand-held control box.

The selection and development led to:
• The elimination of the manual handling hazard.
• The unit allowed an estimated increase in productivity of 25-30% with an actual increase in weld quality due to the consistency of the weld.

Case example

The development of business banking facilities

As a mutual organisation that existed for the benefit of its members, an institution had never embraced business customers, as it essentially catered for the needs of individuals. Corporatisation, however brought new pressures and the expectation that it would perform against the same benchmarks as its competitors. To this end, an unwritten condition of its licence to operate as a corporate body, was the requirement to develop products that would satisfy the banking needs of small- to medium-sized businesses. That is, business banking was to be developed.

The major constraints were:
• Time – To satisfy the various external interests, time was of the essence in the CEO's mandate to implement business banking. However, due to the lack of a formal project method, this requirement actually caused undue haste, re-work and uncertainty in the project requirements.
• Electronic delivery – No matter what product was delivered to customers, it had to be delivered electronically, so as not to interfere with existing operations. This was later proven to be unrealistic, and both old and new customers were served through a branch network – although sophisticated electronic delivery options were available.

The absence of cost and performance constraints contributed to the delivery problems.

Case example

Introduction of an electronic appliance – time constraint

The project involved bringing a new electronic frypan onto the local appliances market.

The owner required the frypan to be on the market after nine months. This decision was a marketing one, and so the owner employed specialists in the field of frypans and controllers to ensure the completion date. It was not until the project meetings began with the specialists, that the owner realised that its completion date would not be met due to the long lead times of several electronic components, and the long development time of the dies for the frypan base and lid. This was one key issue that the owner did not fully analyse.

The owner had initially thought that it could complete the project within a nine month period. This time frame was more or less 'plucked out of the air', as there were no supporting documents such as a timeline, no concept of research and design restrictions to time, and little knowledge of the lead times of several components and processes involved in making an electronically controlled frypan.

EXERCISES

1. Does it really matter if people don't know the difference between a project and an end-product, from an applied project management viewpoint? As a side comment, it is remarked that most people, including many senior project managers, don't know the difference. Discuss.

2. Does it really matter if people don't know the difference between a project objective and an end-product objective, from an applied project management viewpoint? As a side comment, it is remarked that most people, including many senior project managers, don't know the difference. Discuss.

3. What is wrong with stating objectives, either for a project or its end-product, in the following forms:
• To satisfy a political undertaking.
• To demonstrate a government commitment to development.
• To reduce soil erosion.
• To ensure availability and continuity of funds.
• To keep the community informed of progress and expected outcomes.
• To reduce user operating costs and times.
• To increase user safety.
• To reduce maintenance costs.
• To increase load carrying capacity.

These examples show a common way that objectives are stated. Does it really matter if people don't know how to state objectives properly, from an applied project management viewpoint? As a side comment, it is remarked that most people, including many senior project managers, don't know how to state objectives properly. Discuss.

In answering this exercise, think in terms of the non-uniqueness of the solution that results where the objective is not capable of being extremised.

4. How realistic are we when we choose weighting functions? Can these weighting functions be substantiated?

5. Suggest some typical project objectives (other than those given above).

6. Suggest some typical project constraints (other than those given above).

CHAPTER 8

Project Scope

8.1 SCOPE DEFINITION

Scope is what is involved in undertaking the project, the extent of the work to be undertaken, what work is contained in the project (perhaps by defining what work is not included in the project, in effect the boundaries to the project) that leads to the end-product. Scope is fully described by listing all the project activities.

There is much confusion over the usage of the term scope. Some writers wrongly include the end-product, resources, quality standards, performance measurement, acceptance criteria, assumptions, approach or methodology to be used, justification, approving body, cost implications of options, timing, environmental (natural) impact, employment considerations, political ramifications etc in the definition. While useful information to have, it is not scope, and should be clearly distinguished from scope. Some see a scope statement as a description of deliverables (end-product plus state – See Chapter 16 – final (terminal) conditions for the project) or 'what the customer expects to see as a result of the project'. The process and the result are confused. It appears that the word scope is used by most people in the sense of Alice in Wonderland:

> 'When I use a word,' Humpty Dumpty said in a rather scornful tone, 'it means just what I choose it to mean, neither more nor less'.
> 'The question is', said Alice, 'whether you can make words mean so many different things'.
>
> ('Through the Looking Glass', Ch. 6, Lewis Carroll)

The biggest transgression though is the sloppy, interchangeable use of the terms 'objective', 'constraint' and 'scope'; few people appreciate the distinction and why there is a distinction.

Commonly the carriage is put before the horse, such that the scope determines the objectives and the constraints. This confusion and imprecision stifles the understanding of project management.

8.2 GETTING TO THE SCOPE

Getting to the scope is the process of working through the inverse problem associated with the 'means to the end-product' of Figure 5.1, Chapter 5. Objectives and constraints are delineated, alternative work methods are proposed and evaluated against these objectives and constraints, and a best work method selected. This defines the extent of the work or scope.

[Planning is an associated inverse problem that depends on first having defined the project objectives and constraints. Planning envisages *how the job will be done, in what order* and *when* and *with what resources*.]

As a project progresses, the objectives may change but more likely the constraints will change (for example, the work completed so far constrains what can be done next, assumptions on delivery times were inaccurate, ...). This leads to scope changes. Management is an on-going sequence of decisions (in consultation with interested parties) – solving the inverse problem repeatedly and with new/changing constraints (and possibly new/changing objectives). These decisions are not made using neat mathematical techniques such as linear programming or Pontryagin's maximum principle, but rather involve judgement (individually or collectively), expert or otherwise.

It goes without saying that to get the end-product and scope right, it is first necessary to identify correctly the objectives and constraints, which in turn follow from a correct translation of the owner's and stakeholders' values and requirements. Some organisations use checklists to ensure that the owner's values and requirements are correctly interpreted. Without correct interpretation, the wrong scope and wrong end-product result, and continual revisions result in lost time and frustration in the management of the project.

8.3 SCOPE PRACTICES

It is considered good practice to delineate the project work at the outset, to communicate this to all project stakeholders, and to ensure that the stakeholders understand the scope. This delineation of the project work might be referred to as a *scope statement*, if given in broad terms. [Note, that some writers adopt a broader, but wrong, usage of the term scope statement to include deliverables, objectives, and constraints as well as some justification of the project. They confuse it with an approval document.]

A scope statement may be referred to as a 'statement of work' (SOW), a narrative of the project work.

The extent of the project work follows from a knowledge of the project objectives and constraints. Work established without reference to objectives or constraints could be expected to be non-optimal with respect to the objectives, and also possibly violate the constraints. Some people are able to rationalise this as being satisfactory, and some even rationalise it on the basis that they are being 'practical'. In reality, though, such people are unaware of how the scope follows from a knowledge of the project objectives and constraints.

Attention is paid to scope in order to indirectly control deviations in cost, time and quality from that planned. Scope changes on projects are common and may be referred to as *variations* or *extras*. Some projects have a tendency for multiple changes leading to what is sometimes termed *scope creep*. Scope changes may come about through changed thinking, errors, changed values and circumstances, unforeseen issues etc.

People confuse a change in the end-product with a change in scope; the first will commonly lead to the second, but they are not the same. People also wrongly talk of control when they speak of *change control*; invariably what is involved is a monitoring and recording of changes and perhaps some restrictions on changes. Changes then lead on to replanning etc of the remaining part of the project. True control is in the sense of Figures 8.3 and 16.3.

Scope statements may be present not only for the project but for subprojects. Subproject work, that is to be done by outside contract, requires a scope statement. This is so whether the contract is for services, supply or physical work. Where services are involved, such as consulting services, the scope statement might be called a *brief*. (Additional information might be included in the brief, such as remuneration, and reporting requirements.) Scope statements provide a basis for communication, and perhaps agreement, between the relevant stakeholders or parties involved.

Case example

Building security system

A subcontract project involved the supply of hardware and software for building management, security and photo-ID for a prestige building.

The scope was defined as: the supply of standard photo-ID software and hardware; services to integrate the software and hardware with existing system and systems being installed concurrently; design, coding and testing of specified enhancements to the standard software; systems integration and systems test.

This scope statement was derived directly from the proposal that had been presented to the contractor by the subcontractor.

Scope becomes better delineated as the project progresses. For many projects it is not possible to know the exact scope at the start of a project.

Case example

Road design

An example of better delineation is road design which is affected by vegetation surveys which show significant areas of rare and endangered plant communities with little or no option for deviation of the alignment. Similarly the identification of areas of tree die back disease place significant restraints on methods of construction which in turn affect design.

Some writers feel a need for distinguishing 'scope management' as a core project management function. Mention is also made by such writers of 'scope planning' and ensuring that 'enough, but only enough, work is undertaken to deliver the project's purpose successfully'. However, it is debatable whether such a distinction is necessary if a clear framework is established for what project management is fundamentally doing. For example, planning and control practices directly follow from stating the inverse 'means to the end-product' problem.

Case example

Information system

Background

The senior executive of a university saw the need to overhaul its methods and tools used to manage student-related information.

The university's computing department was asked to consider options available to implement a new information system, one that catered for all facets of maintaining administrative information on students, and integrate this with existing systems (ledgers, accounts receivable, fees etc).

This case study follows the scope issues of this information system development project.

The search for the 'best solution'

As the needs for such a system were quite unique, there was little available in the way of 'off-the-shelf' packages in the marketplace.

A survey of other universities revealed that they too were in need of such a system, and some had formed alliances with other universities to develop systems. It was discovered however that such systems did not perform many of the functions required by the university; as well they lacked the technical capability of processing large volumes of data, a consequence of a large student population.

Following these preliminary investigations, the computing department put forward a proposal to design and develop its own information system, which could also be sold to other universities to fill a gap in the market. The proposal was accepted by the university senior executive and funding was allocated to begin development.

Enter the consultants

An external consultant was employed to define the capabilities, structure and specification of the new system. This was achieved through a process of interviews with the stakeholders, a review of current systems, and the consultant's own expertise in the area.

Upon completion of a definition of the system capabilities, a proposed project program and budget was forwarded to the university senior executive. The project program and budget far exceeded original estimates because the consultant had defined a comprehensive system which catered for all exceptions – a 'rocket ship', when all that was required was a 'car'.

The university senior executive decided that it could not wait the several years estimated, nor could it afford the budget required, and thus only approved approximately half the funding asked for, and set a shorter deadline.

Following the university's decision, a different consultant was hired to redefine the capabilities, structure and specification of the system, and at the same time manage the running of the project; a project manager was born.

The new project manager immediately went about stripping 'non-essential' components from the system and, within weeks had developed a cut down version, which was shown to the stakeholders before development work began.

Contractors were hired to do some of the programming tasks along with internal staff.

Moving goalposts

Throughout the project, one of the major stakeholders, the student administration, kept asking for additional functionality and modifications to the system.

A significant degree of scope creep was becoming evident, and as a result, the project fell behind schedule. The project deadline had been fixed, because the university had committed to using the new information system for enrolments. Hence the project manager was forced to put a freeze on all requests for changes. It was also necessary to further reduce the functionality of the delivered system to the most crucial subsystems to ensure the system was delivered on time. Subsystems that could not be delivered by the original deadline were earmarked for a subsequent project (stage 2).

System delivery

The new information system was delivered just in time for the enrolment period, with staff working up till the night before enrolments to resolve last minute problems. Although the system was now in use, there were still a lot of 'on the fly' fixes required to fix problems because there was not sufficient time available to comprehensively test all modules. The project manager's contract was ended and the computing department took over the new system.

Post instalment

For several years after the commissioning of the new information system, the computing department continued to develop stages 2 and 3. This process took a lot longer than expected because the scope was constantly changing, and was allowed to do so because there were no forthcoming critical deadlines.

New problems were emerging because of a university paradigm shift and departmental restructuring to a 'user-pays' philosophy. What the computing department categorised as new developments, the student administration classified as system maintenance. This caused problems because maintenance was centrally funded, whereas development work was charged to the department requesting it.

Comment

There were many examples of both good and poor practices. The major observation in hindsight is that there was not enough work done in the initiation phase of the project.

The objectives were well defined but not the constraints. Initially, a system was designed that would do everything anyone could ask for, but this was unrealistic due to time and budgetary constraints.

The second consultant was able to provide a system which delivered the basics of what was asked for, in an acceptable time period and budget, albeit minus several modules which could be considered a 'wish list'.

Alternatives were sought in the form of 'off-the-shelf' packages and other university systems, but neither was considered acceptable.

Authorisation

Perhaps the most critical error was the non-signing off of the system capabilities by the project owner, the student administration. The computing department began work on the system without the owner officially approving the capabilities.

This led to the delivery of a system consisting of only the essential components by the deadline. More funds needed to be committed by the university to cover the full extent of the work.

For several years after the delivery of the initial system, the lack of authorisation still caused problems because there was no agreement on what was to be delivered. This issue manifested itself in squabbles over whether a module was part of the initial project (and therefore defined as system maintenance) or a new development, thus requiring additional funding from the project owner.

Scope statement

The capabilities of the system to be developed were not as clearly defined as they could have been, and the scope of the project itself, with its constraints and limitations, was not thought about until the deadlines drew near and funding began to dry up.

The computing department delivered a system that it thought was what was required. The computing department should have insisted that the system capabilities were defined and signed off by the stakeholders. Had this been the case, the computing department would have known what was required and planned accordingly from the outset. There would have been fewer excuses available to project staff for not attaining targets.

Scope changes

In the early days of system development, the limiting of scope changes was very poor. The project stakeholders were all asking for system changes and additions, and for the most part they were included. As the project neared the crucial deadline, the project manager rightly froze any further changes to the system, and hence changes to the scope.

Case example

Industrial in-house projects

It is believed that by far the greatest single factor responsible for poor project performance (consistently exceeding cost and time targets listed in the approval documents) in capital projects undertaken by one company was a lack of definition of scope.

Failure to clearly identify the scope resulted in delays and increased costs which were attributed to five distinct areas:

• Changes in scope to suit refined needs in many cases involved demolition and removal of work completed by the current or a previous project.
• Cost of lost time charged by the contractor while waiting on the owner to revise design drawings, source new equipment, gain approval for additional funds etc.
• Changes to design and scope as a result of the 'preferential engineering' of new personnel.
• Airfreight delivery increased the costs of materials and major equipment items not specified in the original design (a substantial expense in a remote location).
• Construction of significant amounts of 'nice to have' extras.

The lack of detail in the project approval document meant that in the majority of cases, it was often impossible to prove that these modifications and additional work were in any way included in the original scope and estimate submitted for approval.

As a result, the capital expenditure process was overhauled to place greater focus on clearly defining scope, and cost and time targets at the time of submission.

The main problems above in particular were addressed and the project group was charged with the responsibility for ensuring that new procedures and processes were developed to ensure:

• The project submission for approval included sufficient detail about the scope of the proposed project to ensure that major changes could be readily identified as being outside the scope of the approval.
• Delays as a result of changes in scope and design could be eliminated through a clearer understanding of the work involved before the work commenced.
• 'Preferential engineering' was considerably reduced through faster completion of projects, thereby avoiding the impact of a high turnover rate of project management personnel.
• Airfreight costs and associated delays were eliminated.
• Unnecessary extras were eliminated.

Revised procedures specified that every submission for capital expenditure must be signed off by all stakeholders, including those on whom the project would have little impact. While this meant that preparation of a submission now required considerably more time and effort, the results exceeded all expectations.

Apart from greater achievement of cost and time targets, improved definition of scope in the submission document resulted in:

- Greater acceptance and therefore earlier handover of the end-product to maintenance and operations as a result of wider ownership of the design.
- End-products being more 'fit for purpose'.
- Improved efficiency of project management resources as a result of being able to better plan project work.
- Improved safety of construction, operations and maintenance personnel through wider awareness of detailed design and scope.
- Improved management of contractor resources.
- Many projects cancelled as a result of stakeholders becoming more intimate with the project, and selecting an alternative means of achieving a similar result (for example, changes to operational practice).

Case example

Computer systems – steel plant

The broader project involved the development of a steel refinement plant. The information technology subproject involved the development and construction of the computer systems needed to operate the plant. All activities were performed in house.

The scope of the subproject derived from the information technology tender document:

- Perform risk analysis for the integration into existing systems.
- Identify constraints within each system environment.
- Develop and design subsystems for each plant module (planning and scheduling, weighbridge, scrap yard, steel production).
- Integrate plant modules to maximise efficiency and reliability.
- Coordinate all production and development activities in accordance with company quality standards.

Some specific activities included:

- Detailed design, specification and programming on a mainframe of the planning and scheduling subsystem which converted the orders placed on the mill into a production schedule.
- Detailed design, specification and programming of the weighbridge and scrap yard subsystem which recorded the weights of deliveries, despatches, internal movements and tares.
- Detailed design, specification and programming for the steel production subsystem which provided functions in the melt furnace, ladle furnace, caster and billet yards to determine the grade and specification details necessary for heating steel, and to record production details.
- Specification of computer equipment which would remain on site after implementation.

Case example

Truck refrigeration

Background

The subject of truck refrigeration was raised as a result of the following:
• The receipt of letters from large stores listing the future checking of product temperatures at their receival docks.
• Enforcement of quality assurance (QA) requirements.
• Enforcement of a refrigeration code of practice.
• The company's strategic plan.

The plant engineer was asked to investigate the issue of truck refrigeration and produce a report, listing the issues and proposing possible corrections.

The steps taken involved:
• A transport refrigeration specialist was commissioned to evaluate and to provide an estimate for repairs or upgrades on the current truck fleet. A report with suggested upgrades and costs was received within a week of the investigation.
• A questionnaire developed by the plant engineer was sent to each branch manager. The questionnaire consisted of typical temperatures achieved on each truck, temperatures required, typical profile of goods carried, number of customer complaints, satisfaction with current refrigeration and any other comments.
• The result of the questionnaire was collated and tabled by the plant engineer. This table listed the following by branch: number of trucks with inadequate refrigeration, upgrade costs based on this, number of trucks getting customer complaints on temperatures, upgrade costs of trucks which attract customer complaints, trucks unacceptable from a QA point of view, and upgrade costs of trucks unacceptable from a QA point of view.
• Temperature profiles of each truck were measured using a temperature data logger.
• The graphs and data were analysed and a summary presented to the senior management team by the QA adviser. The report consisted of the transport refrigeration specialist's conclusions and recommendations, results of the questionnaire, one of the truck's temperature profile graphs, a proposal to upgrade all the trucks within the company and a proposed three year lease plan.
• A truck fleet refrigeration plan, listing the actions, responsible person(s) and dates, was issued by the corporate services manager after the investigation was conducted.

The team members

The team consisted of the plant engineer (project manager), an external specialist, the QA adviser and a branch manager. Membership was on an 'as needs' basis.

Comment

The project was 'doomed' from the beginning, predestined to become confusing as none of the activities that are normally carried out early in a project were discussed and documented.

Activities such as the establishment of objectives, constraints and the scope of the project would have aided in the planning and data gathering.

The project began as a reaction rather than as a planned and agreed need as identified by the management team.

The project was not defined, and so it followed that both the stakeholders and project team members all had different perceptions of what the project was about.

The statement 'investigate the issue of truck refrigeration' was interpreted as 'find out what is wrong with the truck fleet refrigeration and make recommendations to fix it'.

In hindsight, the project would have been better defined as: to review the current status of refrigerated trucks by evaluating against the listed criteria and definitions. Examples of criteria – compliance with the refrigeration code of practice; temperature profiles of each refrigeration unit; and whether the units were independent of the truck's engine, capacity, and body assessment.

The scope was not defined as can be seen by the various boundary changes within the project – the transport refrigeration specialist evaluated the fleet at one branch; the questionnaire was forwarded to all branches; and the investigation was limited to the performance of the refrigeration units only although the body of the truck would have an effect on the results.

The transport refrigeration specialist was given a verbal (rather than a written) brief by the project manager and this was reflected by a prompt response and concise information within the agreed price.

The project end-product was not specified, but was implied as a report. The same applied to the final acceptance criteria and the approving body; these were not specified. Scope was allowed to creep. At the conclusion of the truck refrigeration project a recommendation was made to upgrade all current trucks at every branch based on a limited study at one branch only.

Case example

Selection of a faxing system

Senior management wanted to leverage their investment in PC computers and a local area network to provide a faxing capability for the office personnel.

Scope confusion

The scope (for the task of selecting a preferred faxing system) was defined as '*PCs running a defined operating system*'. Explicitly excluded were all other computers and operating systems. Clearly the use of the term 'scope' here is confused with the end-product. This is not uncommonly done by project people.

A reworked scope

On pointing this out, the scope was reworked to the following:
- *'Ability for the user to send and receive faxes.*
- *Ability to send a fax from any PC based application that one is normally able to print from.*
- *The product must be easy to use; thereby minimising the user training costs. For an average user, it should take less than 30 minutes tor training.*
- *The product should contain good system administration tools; thereby enabling the help desk to be able to quickly and efficiently help users.*
- *The vendor must offer high quality technical support.*
- *Problems (hardware and/or software) must be fixed within 4 hours of being reported.*
- *Ability to use this product on PCs running a defined operating system.*
- *Ability to use this product on other computers and operating systems was not required.'*

Clearly this is also not the project scope, but still refers to the end-product. Confusion still exists over what 'scope' actually means.

EXERCISES

1. Examine a contract to which you have access. How does that contract define scope? Is this definition consistent with the comments here?

2. List what you believe are the main reasons why the scope of a project changes during its lifetime.
   Have you ever been involved with a project in which the scope did not change? What were the particular circumstances that enabled this to happen?

3. Some people include the project objectives in a scope statement. A different view is presented here. Which approach do you prefer? Or doesn't it matter? Justify your view.

4. If you are a consultant, or deal with consultants, examine a consultant's brief to which you have access. What else besides a scope statement does it contain?

5. Scope management is not something that many project people think about explicitly. Most people commonly only talk of time, cost and quality. What value do you see in identifying a group of activities under this banner?

PART C

OTHER LIFE CYCLE ACTIVITIES

CHAPTER 9

Other Life Cycle Activities

9.1 INTRODUCTION

This chapter gives a summary of the activities remaining in the life cycle of a project. Such activities constitute the bulk of the many available project management texts, and hence are deliberately only mentioned in passing in this book. Activities are listed according to:
- The remainder of those things that are peculiar to starting off a project, and generally don't occur again in the project.
- The recurring issues within the body of projects. They give rise to synthesis problems with ever increasing (evolving) detail available.
- Those things that are peculiar to finishing off a project and generally don't occur earlier in projects.

9.2 OTHER ACTIVITIES IN STARTING A PROJECT OFF

In summary, other activities involved in starting a project off include:
- *Owner involvement*
  [In-house versus outsourcing/contract decisions (Carmichael, 2000); Joint ventures, ...]

- *The project manager*
  [Selection and appointment; Contract for services; Qualities and characteristics, responsibilities; Project management delivery method (Carmichael, 2000); Project management services contracts; Legal matters, liabilities, ...]

- *The potential project team*
  [Looking ahead to team building.]

- *The project stakeholders*
  [Identification and management; Project owner; Project manager; Potential project team; Community; Community consultation; Authorities; ...]

- *Support studies*
  [Data collection; Site inspection; Support studies – site/locale, market, risk/uncertainty and assessment, environmental (natural); Standards and regulations; ...]

- *Alternatives*
  [Generation of alternative end-products and means to end-products; ...]

- *Approximate estimates*
  [Approximate estimates of: resources, money (funding, income), times (schedules)]

- *Feasibility*
  [Technical and economic feasibility; Approximate estimates of: resources, money (funding, income), times (schedules); Pre-feasibility studies; Feasibility studies; Risk analysis; Economic evaluation; Non-economic issues, ...]

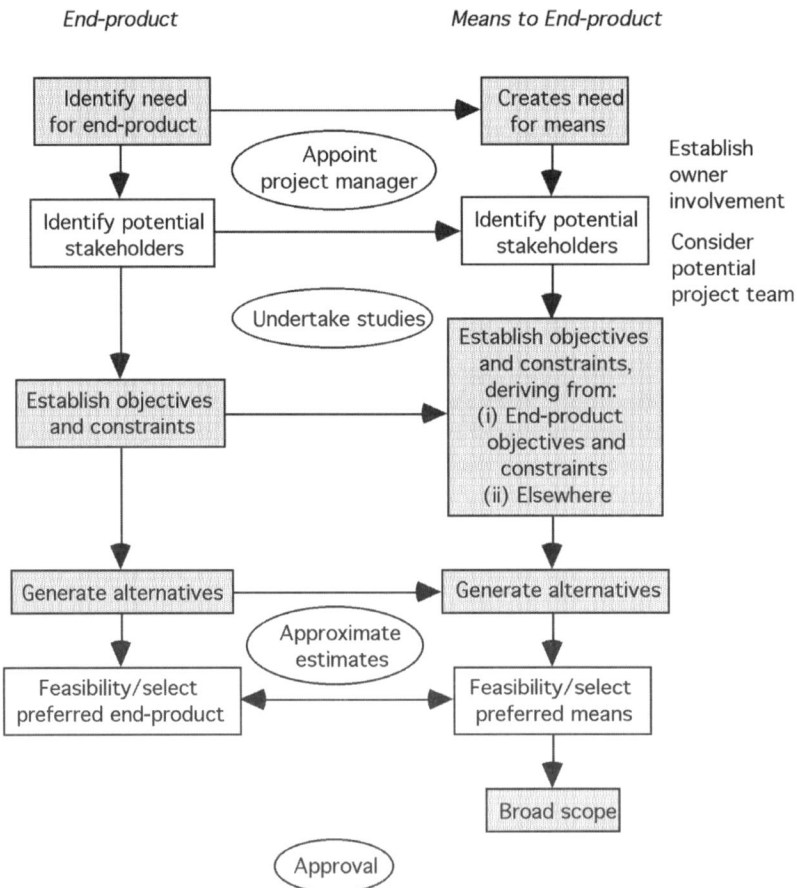

Figure 9.1 Project initiation activities.

- *Approvals and authorisation*
  [Summary report; At the end of the project initiation work, present a summary report of findings and recommendations. Approval to proceed further could be expected to depend on this; Authorisation; Obtain approval from – Statutory authorities, owner, ... Often the transition between project stages.]

The place where these issues occur in the initiation of a project are shown unshaded in Figure 9.1. The shaded portions refer to Chapters 5 to 8.

9.3 MAIN ISSUES IN PROGRESSING A PROJECT

In summary, the main issues involved in progressing a project include:
- *Planning and controlling the work, finances*
  [Scope; Work packages; Definitive and detailed estimates of resources, money (costs, funding, income) and times (schedules); Planning (schedule, budget, cash flow, resources); Monitoring; Reporting; Controlling; Performance measures; ...]

- *People issues*
  [Project team, responsibilities; Organisational structures; Team building; Motivation, leadership, delegation; ...]

- *Materials and equipment*
  [Procurement; Management; ...]

- *Procuring goods and services*
  [Outsourcing; Project delivery methods (Carmichael, 2000); Contracts; Contract payment types (Carmichael, 2000); Tendering; tendering and contract documents; Materials and equipment procurement; Contract administration (Carmichael, 2002); Disputes (Carmichael, 2002); ...]

- *Handling information*
  [Communication; Management; Documentation; Meetings; Policies and procedures etc; Reports; ...]

- *Workmanship and standards*
  [Quality; Management; ...]

- *Uncertainties*
  [Risk; Management; ...]

- *Resolving problems*
  [Including value management studies; People, technical, contractual, ... problems]

- *Approvals and authorisation*
  [Summary reports; At milestones, present summary reports of progress. Approval to proceed further could be expected to depend on this; Authorisation; Often the transition between project stages.]

In parallel with these issues, the technical practices of *design/engineering* and *construction/fabrication/erection/...* would be proceeding.

Figure 9.2 indicates the interconnectedness of all these issues. It is a non-sequential process, with issues being revisited often or continually.

9.4 MAIN ISSUES IN FINALISING A PROJECT

In summary, the main issues involved in finalising a project include:
- Commissioning.
- Asset management activities.
- Finalising accounts.
- Data compilation.
- Releasing resources.

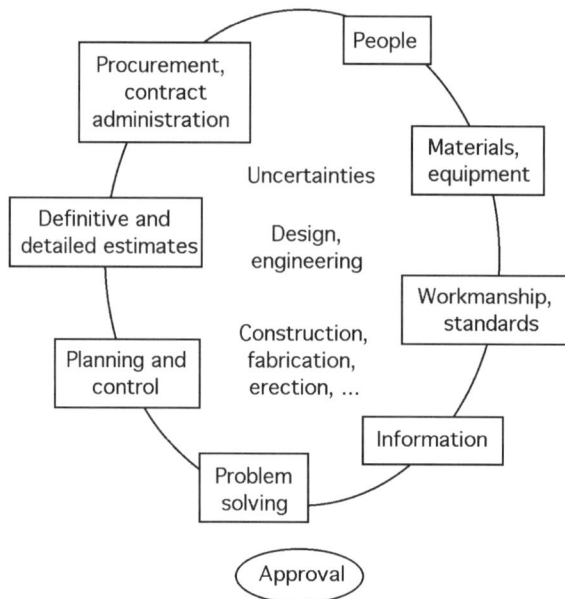

Figure 9.2 Project development and implementation issues.

Figure 9.3 Project termination issues.

- Project completion report.
- Handover.
- Project review; Learning.

The interconnectedness and the flow of issue activity involved in finishing off a project are shown in Figure 9.3.

PART D

PROJECT EXAMPLES

CHAPTER 10

Movie Making

10.1 INTRODUCTION

Movie making (including the making of television shows) is big business. Delivering a movie requires considerable management skills. Interestingly, managing the production of a movie is not generally considered to be project management.

Project manager, in the context of the movie industry, is possibly most closely approximated by the titles and key positions of 'producer' and 'director'. In the movie industry there seems to be variation and overlap between the roles of producer and director. There are cases of producers who direct and directors who produce. Defining roles and responsibilities associated with the conventional title project manager is difficult. However, of these two movie industry titles, the producer typically carries out the duties analogous to a project manager for the entire movie project. A proliferation of titles (for example, the use of the term production manager) in the movie industry doesn't help an outsider trying to understand who has the overall project management role.

Whilst the director is often the person orchestrating the action being filmed, guiding technicians and bringing the film in on time and budget, this is usually only associated with the shooting phase of a movie project. It is the producer who is more of a generalist, staying with the project/movie making through all phases. This person is responsible for taking a film idea and turning it into a motion picture. The producer's job begins long before the start of production and does not end until long after the film is 'in the can'. S/he guides and organises the entire production, from script writing, financing, director/actor/crew hire, shooting, editing and marketing/distribution. The producer can be responsible for a range of different tasks depending on the movie making management structure. Her/his role is to guide the activities through all phases, into a successful motion picture achieving maximum quality at a minimum price tag

The director is commonly defined as the person responsible for all activity in front of the camera, and the producer for all activity behind the camera.

The producer is not the same as the executive producer, a representative, on hand during production, of the film financier.

10.2 PROJECT PHASES

Movie making effort can be divided into phases where a phase transition represents the completion of activities/tasks, or relates to decision points, particularly whether or not to proceed to the next phase.

Possible movie project phasing terminology is:
- Initiation.
- Pre-production.
- Production.
- Post-production.

The phases loosely reflect a generic chronological model for project management. In the initiation phase, the decision to go forward with the film making would be made – there is no point in proceeding to conduct pre-production activities if there is not a good film idea or funding is not secured.

There may be overlap between phases, and the making of different films, depending on their origin and nature, may interchange phases in which certain activities are carried out.

Initiation phase

Other terminology for this phase might be start-up, development or feasibility phase. This is the concept phase. A stakeholder has decided that a movie could be made, possibly based on a draft script or concept. This phase establishes whether or not it fills a market need and a film is feasible. Both project and end-product objectives would be defined. Scope, budget and schedule are broadly established.

Key activities undertaken during this phase might be:
- Conceptualisation. A stakeholder has an idea for a film, or a script is selected. A broad outline of the story is decided and usually thumbnail sketches or storyboards are drawn to convey plot and to aid with visualisation.
- Consideration of possible producers and directors (if this has not already been decided).
- Consideration of possible casting in the principal roles. Lead actor commitment may occur in this phase or the following phase. Movies may be better 'sold' if crowd-drawing lead actors agree to roles in the film.
- Considerations of where and how the film is to be made, for example, on location, in a studio, whether to use special effects or animation.
- Making estimates of costs and time. Budget preparation – the overall budget required to produce the film is established. A preliminary program is made.
- Funding/financial backing search. The concept for the film and storyboards are used to pitch the idea to a production company, film studio, bank or non-profit film association that supports the arts, or financial backer/partner, who may be able to provide funding for the making of the film.
- An assessment of feasibility and risks. The end value and market of the end-product may be hard to establish up front.

Progress to the next phase depends on funding availability and feasibility.

Pre-production phase

This is the planning and development phase, where preparation is made to shoot the film. Funding arrangements are confirmed and formal and legal approvals sought with relevant authorities. The financial backers may require any risk to be reduced prior to fully committing resources. Key team members are appointed and an organisational chart produced while the script and film making schedule are developed. The schedule reflects the budget, the time a director needs to shoot, and the logistical problems of the script.

This is an important phase of the process where the future success of the movie production is determined.

Key activities undertaken during this phase might be:
- Script writing – the film concept may be converted to a script by a writer, if not already done so. The screen play is developed.
- Commence casting the roles.
- Actors start rehearsals and the script may be reworked as a result.
- Hiring of production crew.
- Hiring of music composer and composition of musical score.
- Suitable locations for filming are investigated. Studios are selected.
- Technology review.
- Hiring of equipment required to make the film. Preferred suppliers identified.
- Production of sets and costumes commences.
- The logistics of supporting, accommodating and feeding the cast and crew is addressed.
- Lighting design.
- Choreography.
- Cinematography plan – camera angles, focus, close-ups, moving, stationary etc.
- Rehearsals and preparation – rehearsals may result in areas of the script being altered. This includes any dance and musical rehearsals and preparation for stunt work. Make up and hair conceptualised and trialed.
- Shooting schedule, which outlines when each scene is to be shot, locations, equipment and set requirements, and cast and crew required for each day of production.
- Production companies are selected.
- Budgets and schedules are refined, managed, and controlled. The total film making budget is allocated to all areas of the film making process. Areas include wages, equipment hire, set construction, costumes, venue hire, catering, accommodation, marketing and distribution. Payroll organisation.
- Source insurance.
- Marketing plan and merchandising plan; determine audience and plan advertising for that audience. Select distributors.

When all is ready, production or shooting can commence.

Production phase

This is the main execution phase of the project, where there are intense, parallel activities by many groups, and major expenditure. The producer is responsible for a problem-free shoot, and for handling both the logistics and the overall organisation.

Key activities undertaken during this phase might be:
- Location and studio shooting are conducted.
- Supporting animation and special effects commence.
- Accommodating and catering for cast and crew.
- Budget, schedule and quality monitoring and control. Daily production reports.

The output of this phase is many reels of film ready for editing.

Post-production phase

The final product is the completed movie ready for distribution.

This phase is important for the movie's financial success as it incorporates the marketing and sale of the production to an audience. In this phase, the likely success of the movie in monetary terms can begin to be measured by initial market testing and then box office takings.

There is an evaluation and documentation of the project in this phase, with a view to future projects, to avoid repeating costly project management pitfalls such as time delays and cost over-runs, and to record efficiencies. Traditional movie making may not emphasise this.

Key activities undertaken during this phase might be:
- The movie is edited. Film shoots are put together to form a coherent continuous film, including the incorporation of digital special effects.
- Re-shooting as required and as the budget permits.
- The music score is finalised and matched to the footage.
- Sound. Voice overs or dubbing as required, sound effects, and finally sound mixing (the smooth combination of all these sounds). Re-dubbing sections of poor sound in the studio.
- Marketing/promotion. Prior to film release, 'opening nights', continued promotion.
- Film release.
- Film distribution – in theatres and plans for video and television release dates, and re-runs.
- If the movie proves a success, further money can be extracted from the public through merchandising.

10.3 PROJECT MANAGEMENT

Even for shoe-string budget films, there are parallels with what is understood to be project management. A producer may need to be a combination of shrewd business person, tough taskmaster, prudent cost accountant, flexible diplomat, and creative visionary, giving motivation, understanding and coordination to a group of highly technical and highly artistic people, in many cases using leading edge technology. Good communication ensures the smooth production of the film within the time frame allowed. Scope is partly dependent on the script, but can change with the addition of extra scenes or alternative scenes.

Money

Managing production costs to fit a budget is important. For all film making, a budget is allocated by a production studio, bank, film institute, ... Some costs, such as director's, producer's and actor's salaries, are fixed. Some costs, such as sets, travel, transport, equipment hire, crew and salaries, catering, distribution and music rights, are not fixed. If the producer runs out of money before the end of filming then no film will eventuate and no return on investment will be received. Thus expenditure is carefully monitored and controlled. Contingency planning may be part of film making, and funding may be allowed for the unexpected.

Popular film making appears predominantly a business about making money. Effectively managing costs during the project increases the profit, but the success of any film is determined largely by how much money is made at the box office.

Schedule

In popular film making, time is money. The longer a movie takes to produce, the more expensive it is likely to be. Meeting a schedule directly influences the ability to meet budget. Many of the costs associated with subcontractors, location fees, support equipment, ... are a function of time. Production delays due to disruptive weather, poor light, casting conflicts, technology troubles etc drive costs up. Creative wishes need balancing with financial realities in order to avoid costly delays.

Often actors and key production personnel are only available to work on a project for limited time windows. The shooting schedule is carefully planned to ensure individual availability is catered for and then this shooting schedule is adhered to or replanned as soon as timings slip.

People

Movie making is manpower intensive. Actors, the quality of direction and the cast and the behind-the-scenes crew working together as a team can determine the success a film.

Actors may have to be treated sensitively to ensure their cooperation and best performance. The falling out of lead actors with the director or producer may result in the loss of time and money. The production crew and cast work closely together, as does the post-production crew work closely with the producer and director to ensure the overall movie conveys the intended message. Thus a great deal of 'hearts and minds' work is done by the producer and director to ensure the cast and crew, with possibly varying multicultural backgrounds and temperaments, operate as a team. The end-product reflects how people interact on screen and behind the scenes, and the working relationships developed on the project.

Quality

The box office determines how successful the movie will be, and therefore the return the movie makes.

In the pre-production phase, the hiring of quality equipment and skilled actors and crew may largely determine the quality outcomes. If cut rate or unskilled workers, or B-grade actors are hired, then there is little that can be done during production to improve the quality of the output. Film quality is also dictated by securing adequate funding for production, so that corners do not have to be cut in set construction, costumes, special effects, editing etc.

Risk

Given the artistic nature of film production and the many things that could go wrong (actor-director squabbles or injuries, production overtime etc) producers need contingency planning. Little effort seems to be put into risk event identification, which means risk reduction measures are not often incorporated – it seems more a case of having money on the side to be used when a risk event occurs. Risks are more likely to be accepted, or avoided through insurance.

10.4 ORGANISATION STRUCTURE

To facilitate the effective project management of movie making (an isolated project), a task force style organisational structure, where levels of authority and responsibility are clearly defined, would be preferred. It provides for easier budgeting and cost control.

For making movie 'M', a possible organisational hierarchy might be as follows. Further lower levels delineate the roles or job titles of individuals. The size of the project team can be very large, with specialists in every position.

M1   Executive Producer
M1   Producer
   M1.1   Director
      M1.1.1   Production management
      M1.1.2   Acting
         M1.1.2.1   Casting
         M1.1.2.2   Actors, lead, support
         M1.1.2.3   Stunts
         M1.1.2.4   Dancers
      M1.1.3   Photography
         M1.1.3.1   Cameras
         M1.1.3.2   Stills

	M1.1.3.3	Filming
M1.1.4	Lighting	
M1.1.5	Sound	
	M1.1.5.1	Music, songs
	M1.1.5.2	Sound effects
M1.1.6	Art	
	M1.1.6.1	Sets, design and construction, decorators
	M1.1.6.2	Costumes
	M1.1.6.3	Make up, hair
	M1.1.6.4	Dance
M1.1.7	Technology	
	M1.1.7.1	Software
	M1.1.7.2	Hardware
	M1.1.7.3	Systems support
	M1.1.7.4	Special effects, visual effects
	M1.1.7.5	Animation
	M1.1.7.6	Models
M1.1.8	Story development	
	M1.1.8.1	Script
	M1.1.8.2	Screenplay
M1.1.9	Administration support	
	M1.1.9.1	Property
	M1.1.9.2	Catering
	M1.1.9.3	Transport, accommodation
	M1.1.9.4	Accounts
	M1.1.9.5	Marketing, publicity
	M1.1.9.6	Distribution
	M1.1.9.7	Medical
M1.1.10	Editing	

10.5 FORMAL PROJECT MANAGEMENT PRACTICES APPLIED TO MOVIE MAKING

Formal project management practices don't seem to be used in the production of films, though movie making appears to be based on principles that follow a similar path to a formally managed project. Some suggested reasons for why formal project management practices don't seem to be used are:

• Film making is seen as an artistic endeavour (something that is mentally stimulating, visually appealing and humorous or dramatic depending on the type of story) that should not be restricted by a set, mechanical, rigid management framework. Strict adherence to process rather than form may stifle artistic creation. There is a conflict between art and commerce; the producer is torn between two worlds, perhaps preferring the former and leaving the latter to administrators.

- Many producers do not receive formal training on how to produce movies, learning instead on-the-job and from experience gained while working in lower crew jobs.

- A well managed movie making process or a quality production team or cast is not a guarantor of end-product success. For example, many cult movie successes are movies of somewhat dubious quality with act-as-you-go scripts. On the other hand, large budgeted and well managed film making can lead to box office flops. A movie whose making runs over time and over budget after two lead actors have been replaced can still be a financial and artistic success.

- Movie making is a specialised field that has evolved its own methodology, that may be perceived to be a unique formula for implementing a movie project.

- Formal project management practices have still not achieved saturation in more conventional technology and engineering projects. Many poor management practices still exist in these areas. Growth has been slow. Project management has yet to embed itself in many industries and movie making is one of these.

Movie making is a legitimate project – it follows a phased life cycle, requiring improved definition as it goes; it requires a variety of resources; and it has constraints, objectives etc as do the more traditionally recognised projects.

Project management could assist in the making of movies by providing a management structure, even though each movie is very different in its requirements. It would have particular use for eliminating schedule and cost overruns, in the early recognition of problem areas, in the creation of a shooting schedule, levelling resources and manpower, highlighting critical events in the schedule, reprogramming when delays or changes occur, and the identification of risk events, their analysis and the implementation of risk minimisation techniques.

Project management practices should ensure more efficient use of time and resources and help to ensure the motion picture achieves success, however that has been defined.

EXERCISES

1. Is it only a matter of time before the practices formally employed in project management are more widely adopted in movie production?

2. If project management practices are adopted, would there be a need to adjust the processes to allow for the unique creative nature of movie production?

3. What industry specific drivers will force the transition over time to formal project management practices in movie production? What role will the fact that the producer is ultimately responsible for the end-product play in this transition? With schedule and cost overruns common in movie making, will producers and studios think differently about project management in the future?

4. Is it better to identify risk events and attempt to minimise risk rather than simply set aside funding to address risk events as they occur?

5. Although those within the movie industry may see the industry as unique, is applying formal project management practices to movie making feasible? Would it be any more difficult than applying these practices within the traditional project arena?

CHAPTER 11

Organisational Change

11.1 OUTLINE

Organisational change involves the re-structuring of processes, culture or relationships over part or all of an organisation. Change may be externally driven or forced, through factors such as technology advances, competition, changing regulations or economic conditions, or internally driven, in the belief (correctly or incorrectly) that the present situation can be improved or needs improving.

It is almost impossible, a priori, to state whether an unforced change will bring about improvements or a degradation, or even that the cost associated with a change is justified. As such, unforced change is usually undertaken on the basis of a hope of improvement, but sold with much fervour and rhetoric, particularly by senior management trying to justify their positions, management consultants looking to generate work for themselves, and by writers of popular management paperbacks. In order to feed these three groups of people, change has become more prevalent than it need be, and change management is promoted as a separate management discipline. Senior management argue that it is dangerous to be satisfied with the status quo, and they must be constantly doing something in the guise of being prepared for potential external threats and changing markets. In some organisations, there is a never-ending process of change.

Organisational change can be interpreted as a project, and project management principles applied.

11.2 REASONS FOR CHANGE

Numerous reasons (change 'triggers'), some genuine, some bogus, are given for change, and include:
- Movements in technology allow productivity improvements. Examples include new computer software and new machinery.
- Globalisation, political upheaval and a changing world economy.
- Changes in legislation and regulations, may prevent existing practices from continuing. An example is occupational health and safety regulations.
- Operational changes within an organisation. Examples include a new process line, new market, and converting from a non-project-driven organisation to a project-driven organisation.

- Hierarchical changes within an organisation. Examples include downsizing, rightsizing, flattening an organisation's structure, takeovers, and quality related management fads. (Carmichael, 1996)
- Voluntary and forced changes involving employees and workforce relations. An example is conditions for employees, driven by industrial relations issues.
- To improve an organisation's competitiveness, or for survival of the organisation. The above examples may all have competitiveness and survival as underlying motives.
- Change for change sake. Because senior management or a management consultant proposes change. Efficiency may be confused with economy. The view that 'change is inevitable' has its logic confused such that natural change gets 'beaten to the gun' by forced change (in the guise that it is the same as the natural change, only being pre-empted). (Carmichael, 2002, pp. 163-179)

11.3 ORGANISATIONAL ISSUES

Organisational change may be managed using project management principles. For this, organisational change requires defined initial and final (terminal) points – a point where thinking about change starts, and a point at which the change or most of the change is considered complete. For a smooth transition from one organisational state to another, there are a number of factors, related to the individual, the group, and the task, that require addressing.

The individual

People are at the centre of organisational change. Generally people are anxious and fearful about change (both its nature and magnitude), and are concerned with how it will affect them. Without direct involvement in the change process, of the people affected by the change, and commitment (rather than compliance), any desired organisational improvements may not come about. Changes that are driven from the top may not succeed. Commitment comes from involvement and consultation. There is a sort of Hawthorne Effect at work here influencing human and social behaviour. As well, issues of people's needs of remuneration, status, and so on, as well as empowerment and trust, need to be addressed and satisfied.

The group

Each change project could be considered unique. Each workplace is unique. Each workplace has its own peculiar culture. Each person has different wants and needs. Change may alter an existing stable culture. A new culture may or may not be sustainable and stable. The whole organisation needs to support the process. Leadership and communication are important. The project manager needs people skills. Teams and teamwork may be disrupted; new teamwork requires motivation.

The task

Physical changes may need to be made to the way work is done or the business operated. This may involve a re-arrangement of resources and routines. Adequate resourcing is required to install new routines.

11.4 PROJECT MANAGEMENT PHASING

Change may be considered to go through definable project phases.

Initiation

There would be a perceived need for change based on one or more of the above reasons. The project objectives, constraints and scope follow. Responsibility to develop and implement the changes is delegated to a project manager. Alternatives, costs, legal implications etc of the change are considered. Advice and expertise is sought to develop plans and policies to implement the change.

Development

A project team, possibly comprising external consultants and key people who will be involved in implementing and maintaining the change, is organised. Lines of authority are established. The scope is better defined. Responsibility for people with professional or technical expertise is defined.

The change process is planned, outlining the key activities and expected durations, resources and the approvals and hold points. The plan of action is approved by senior management.

Implementation

Senior management support in terms of finance, and the development of new policies and procedures progress the change. The status of the project is monitored and reported with reference to the approved plan of action. Scope changes, quality and costs are also reported.

Termination

The termination of the project can be defined as whenever convenient. For example, it may be defined as when the new operations or procedures are established, even though there may be an ongoing responsibility for maintenance. This occurs with new safety regulations, for instance, where ongoing site safety inspections and new safety plans, policies and procedures have to be maintained.

11.5 PROBLEMS DEALING WITH CHANGE

A common view is that change is not managed well, but rather grappled with in a rather ad hoc fashion, flitting from resolving one trouble to resolving another, much like 'crisis management'. Sensible management principles go largely ignored. A number of problems require addressing with change:

- Teams may adopt their own values and goals, divergent from those of the organisation as a whole. This may lead to intra-organisational conflict and communication problems.
- Being introduced 'cold' to change may overawe people. A gradual introduction to the notion of change, perhaps with a pilot project, may be necessary. Prior consultation may enable unworkable ideas to be rejected, and workable ideas to be accepted. There may be an under-communication of the overall intent.
- There may be resistance to change. By identifying the resistance to change, and dealing with it, the change process may be easier to implement. The reasons for the changes are explained because compliance without commitment may not lead to a real change.
- Older staff are seen as those who may resist the change the most. 'We have always done it this way.' Competition and technology change the work environment and the 'old' methods may no longer be suitable.
- There is a need to communicate the process continually (for example, weekly update meetings or reports) with those affected by the change by providing information about the process and what progress has been made, and to be sensitive to people's feelings. Any job reduction could be expected to include counselling, and career and financial advice for those that are affected.
- Training of all personnel in the change process. Possible personnel development (training). Training and development of staff affected would constitute part of installing the change. This is additional to staff involvement in the change process.
- The change process is influenced by the style and taste of the person overseeing the change.
- Minor changes may still best be able to be handled with formal project management practices, rather than by more expedient memos, announcements and meetings. Over documentation for small changes is to be avoided.
- Band aid and quick fix solutions, perhaps using contracted consultants, are to be avoided. The practice may miss critical issues, particularly related to culture, and work ethics and attitude.
- Monitoring of decisions and progress. Excessive expense.
- Minimising disruption to normal operations.
- Considering the project as being terminated too soon.

11.6 FORMAL PROJECT MANAGEMENT PRACTICES APPLIED TO CHANGE

Formal project management practices appear to be not generally used in managing change. The following reasons are suggested as to why this might be so:
- There is a failure to recognise a change as a unique occurrence with a defined start and finish, but rather view it as a continuous process.

- A general lack of real knowledge of project management practices and what they entail, and a lack of skills. It is easier to go with ad hoc approaches than learn something new.
- A perception that the project management approach may be too inflexible, and more suited to technology style projects rather than people-centred projects.

Project management forces the user to consider issues which may not be considered in ad hoc change management approaches. Risk management reduces the chances of undesirable outcomes. A reactive approach to management is avoided. A structured approach to management leads to less personal stress. People management makes the most effective use of people by identifying, documenting and assigning roles, responsibilities and reporting functions. Communication provides the critical links between people and information that is necessary for success.

The change leader is the project manager who is in a position to work with the people affected in order to make the change happen. This person facilitates the change with minimum disruption and disharmony. Different leadership styles are needed in different circumstances, depending on the nature of the change and the range of people affected. Power, influence and authority issues are important.

Case example

Development and implementation of a new construction cost control software package

This case study relates to the development and implementation of a new software package for managing construction projects. It came about because of technology change.

The contractor undertook a study of construction accounting software to determine the products available and their suitability to its operations. The study was undertaken by the managing director (MD) across a broad range of construction software packages.

After finding none that suited his requirements, the managing director was approached by a consultant to develop software, implement it in the contractor's business and train the contractor's staff in the use of the software. The consultant would later sell the software to others. As all efforts to find suitable software had failed, the MD initiated the project. He defined the scope jointly with the programmers and was also to be the project manager in implementing the change. The project was to be developed and implemented progressively in packages. The change would be regarded as finished upon completion of the software development and its successful implementation into the organisation.

A scope document was drawn up and an agreement made between the consultant and the MD for the project. The MD would have direct influence over the development of the software due to his experience in the construction industry and in cost accounting and forecasting for construction projects. The intent was to have one comprehensive construction cost control package that integrated the accounting, cost accruals, revenue and purchasing areas of the business with the contracting (engineering) side of the business. The software was to be able to account at an individual resource level and later easily return data on overruns and production, as well as all the normal accounting functions.

The MD directed the development of the software. His staff were involved later once each package was operational. With progressive development, the MD's staff could use the relevant areas of the database and then train others and audit the software during trial runs.

The accounting package was the first and most important package, as it set up the data that was required and its entry pro forma. Many standard accounting audits were undertaken as a matter of sound accounting using the database. Although released for use and quality assured by the programmers, many flaws were found by the staff when using the package.

When the engineering side of the package was about to be developed, the MD, in conference with his engineers and the programmers, discussed their various requirements of the accounts and their jobs, and the programmers attempted to develop reports and calculations based on this information. Once again the staff performed a lot of the auditing function, picking up many flaws after the programmers had certified the package as operational.

The MD was not impressed, however his staff were going very well with the training in the new package as a matter of course. Soon the staff knew more than the trainers about the package. The MD took advantage of this and fired the trainers so as to sharpen up the programmers. In the end, the programmers introduced each new package to the staff as complete to the best of their ability, and key operational members of the staff then trial ran the new packages as an owner audit.

The comment was made by all that the MD had incorporated too many resource categories. His own managers agreed and this would mean that to capture the volume of data required by the MD, they would have to employ a number of new people, and change and regulate some administration procedures on the job sites. This would end up costing the company a great deal of money.

The MD was trying to get a comprehensive database, but in some cases it captured too much unnecessary data. Although this was the case, other companies could still use the package without having to go to as great a detail as the contractor. As the package was developed using appropriate technical expertise, it was extremely useful in capturing data and presenting it in the various formats required by the engineers, accountants, purchasing officers and management. The database was then made commercially available.

Questions

1. How did the MD's management style influence the way the change went? Would a management style different from an autocratic one have worked in this case?

Case example

Change issues in a public-sector organisation

To many it appears that there has been a never-ending process of change, an increasing rate of change, that change is almost always forced upon the organisation from the top down with little consultation, and that change has not been managed as well as it could have been.

Within the organisation, the top tier of management's reign is rather finite. Most appointments are for two to three years, and the total number of positions often undergo a significant rotation in a given year. Each leadership group therefore feels that they must implement wide-ranging changes, perhaps in order to be remembered for making an impact. Whilst some change may be necessary, constant reversals of direction and the revisiting of a structure previously occupied undermines worker confidence and commitment. This has happened numerous times, with an example being the disbandment of one group under one manager, and the re-establishment of that same group under the next. It is commonplace for the majority of personnel to be using past names and structures for units that have undergone several iterations since.

Change of this nature is always too compliance-driven from above rather than commitment driven from the staff. Very few changes have been made with consultation, and the workplace has shown little improvement over the short term. Whilst workers simply wish to conduct their activities with the appropriate resources, they find the multitude of levels of management, and the associated bureaucracy, to be stifling. Yet suggested improvements never seem to flow uphill, and the larger organisational changes made (often based on economic, rather than efficiency factors) seem only to restrain the conduct of these same activities. Personnel thus grow despondent and lose their commitment.

Case example

Ore handling; restructured trades supervision

Introduction

This case study describes a change that restructured trades supervision in the maintenance department of a bulk ore handling operation. This change was implemented for the purposes of achieving an improvement in the operation's cost per tonne.

The change was announced as the intention to remove the foreman position across the total maintenance department with a view to implementing some form of self-managed work teams. This basic plan was developed into its final form over a period of approximately two months with the final outcomes being:

- The restructure of the foreman and contracts liaison officer positions into a reduced number of resource coordinator positions.
- The development of the leading hand positions into team leader positions.
- The refocussing of the line mechanical and electrical engineering roles from project-based to reliability and technical support-focused roles.

The total change involved the reduction of approximately twenty trades supervisory positions to approximately ten positions in a department consisting of approximately three hundred total personnel. The change coincided with an offer of voluntary redundancies to achieve the desired reduction, whilst allowing individuals to make a choice about the change.

The project was the restructuring of trades supervision (with the achievement of a reduction in cost per tonne as the business case behind the project). The project selection processes that led to the decision to implement this particular project in order to achieve a cost per tonne improvement are not explored here. All redundancies were required to be voluntary.

Project phases

The project was broken into the following phases:
- Concept.
- Development.
- Implementation.
- Institutionalise.

The breakdown of activities was presented in Gantt chart form showing time frames and activity relationships.

Concept phase
The purpose of the concept phase was to identify the basic requirements, objectives and constraints of the project, and to provide some substance to support a decision about the project progressing. The emphasis was on fully understanding all aspects of the project such that this was an informed decision upon which commitments would be made in the next phase.

The key activities in this phase were:
- Development and documentation of the business case.
- Identification of stakeholders and their various roles required in the project.
- Development and documentation of the critical success factors.
- Documentation of measures of success.
- Development of milestones – including the documentation of the desired maintenance strategy that the change needed to be able to support.
- Pre-feasibility case study research of this type of implementation in other similar industries to identify key issues and learning that could be applied in this case.
- Identification of core elements of the project and expected durations.

- Preliminary scheduling of project activities.
- Approval to proceed to the next phase.

Development phase

Detailed planning was undertaken. Due to this being essentially a human resources change, the design needed to include the people impacted in order to ensure engagement around the plan. Thus there was a heavy emphasis throughout this phase on the engagement of stakeholders. The required activities to be performed in this phase were:

- Appointment of formal project team and clarification of team member roles.
- Engagement of specialist consultants and any additional resources required.
- Design and documentation of the redundancy process.
- Commencing the risk management process for the project.
- Formal communication of the intended change and discussions with other departments, internal maintenance sections and individuals regarding the impact.
- Detailed analysis of existing organisation and maintenance processes to identify:
- Additional changes that would be required as a result of this project, and
- Critical aspects of the organisation that required detailed attention to ensure that they did not get impacted by the change.
- Design and documentation of the new organisation incorporating stakeholder comment (including development of position descriptions, organisation charts, and new or changed maintenance processes).
- Completion of a detailed plan and schedule of changes to be made (including communications plan and breakdown of implementation by responsibility).
- Engagement of key stakeholders and personnel impacted in the final plan.
- Training of personnel for interacting in the new organisation to enable a smooth transition to implementation (for example, teamwork, people skills).
- Setting up and communicating of means of monitoring success.
- Approval to proceed to the next phase.

Implementation phase

The implementation phase was an execution of the plans created. It was where the change actually occurred, rather than being discussed and planned. Due to the constraint of voluntary redundancies, the implementation phase also needed to involve some detailed design which was only possible after final numbers were obtained. Had the basic plan arrived at in the development phase been clearly unattainable after this point, the project would have had to revert to the development phase to consider a new plan. However, the risk management thinking in the development phase catered for the project to continue under most circumstances.

The key activities involved in this phase were:
- Offer of redundancies.
- Determination of final numbers and final format of changes.
- Activation of appropriate risk management responses where required.

- Completion of detailed action plans for each section and person impacted by the change.
- Position specific training (for example, leadership and contract administration).
- Cross-over to new organisation.
- Enactment of detailed action plans.
- Monitoring and correction of minor problems with new structure (monitoring of targets, follow-up discussions etc).
- Fine tuning of position descriptions.
- Fine tuning of maintenance processes.

Institutionalise phase

At the end of the implementation phase, the project was considered essentially complete. The institutionalise phase was the final tidy up. This included the bedding down of cultural changes such that the change became permanent and accepted. The key activities included:

- Completion of corrections to position descriptions and new maintenance processes, and incorporation into documentation.
- Enactment of plans for roles of redundant personnel leading up to their final date.
- Termination of redundant employees.
- Project post-implementation review and documentation of findings (including publishing of findings inside and outside of the organisation as appropriate).
- Formalising of ongoing measures into day-to-day management, considered critical to the ongoing success of the change.

Project management

Issues

The key issues that dominated this project are listed below:

- Designing a new structure without the knowledge of the final numbers and people until well into the implementation phase due to the requirement for voluntary redundancies.
- Applying the same change across the board whilst providing sufficient flexibility to cater for differences between plant sections.
- The need to include a large number of people in the change process, yet achieve a result in a relatively short time frame.
- Managing a potentially wide spectrum of reactions to the change due to the large number of people impacted.
- Managing this change without losing sight of existing operational targets and alongside other organisational changes that were already in progress.
- Managing department morale during and after the change.
- High potential for creating industrial upset.
- High potential for management to lose credibility.
- High potential to ostracise valued employees.

The key themes in this list of issues are the size of the change and the high potential for undesirable outcomes on a broad scale.

Risk
There were a wide range of significant risks that needed to be managed. In particular, there was a high potential for undesirable human resources outcomes, which would inevitably lead to organisational ineffectiveness and ultimately would have cost per tonne impacts. If the risks were not managed, the entire project may well have set the organisation even further backwards from where it was before implementation.

People
Most of the issues touched on some aspect of human resources management; this project was a human resources change to begin with. As well, the specific tasks of changing position descriptions, managing redundancies, and completing job retraining were core aspects of human resources management.

Communications
There were two fundamental department communication channels – internal and external. As the purpose and emphasis of these communications was different, they required different management.

The volume of people and diversity of the audience was high. Different people required different approaches and mechanisms to adequately receive a message. It was critical to the project success that each person received a range of messages correctly and was thus able to participate fully in the change process. If this was not achieved, then the outcomes of other management functions would have been severely stunted.

Money
There were a large number of variables impacting the project's return on investment.

Due to the need for the changes to be made quickly and without impacting on the organisation's effectiveness, it was logical to consider bringing in extra resources to assist with both the day-to-day management and the project implementation. However, this needed to be carefully managed, as the business case for the project was the attainment of a net annual labour cost saving. If it was not carefully managed, the return on investment could have been potentially unacceptable and thus the project a failure. Numerous risk sources were identified influencing the required return on investment.

There was a need to identify the real costs of potentially negative impacts of the project, and the overall impact on asset life cycle costs. Without this information a net reduction in cost per tonne could not be quantified. Due to the number of variables involved and the long time frames, this process was complex.

Case Study

Aluminium production; cost reduction measures

An aluminium production company quite suddenly announced operational cost reduction measures – job losses of approximately 5% – as part of 'repositioning the business for profitable future growth'. These changes were the result of a reduced world price of aluminium, the company's poor return on investment and company share prices that were being out-performed by a competitor.

Project phases

The main phases of the change process were:

Recognition phase
The need for change was identified. The change trigger was the reduction in the price for aluminium, which led to loss of sales and reduced profit.

Assessment phase
A review of the company's business situation was undertaken by investigating customer demand, competition and market situation, in the light of the company's goals. A decision was made to proceed with change.
   Possible remedies for the company were reducing staff, limiting training, reducing production, selling departments/plants, or using new leading technology to improve unit costs.
   The change remedy chosen was a 5% reduction in staff. This reduction process was recognised as not going to be easy. A non-negotiable, autocratic management style was prescribed for the change leader. The reduction was to be based upon employees' past performance and was seen as an opportunity to get rid of 'dead wood'.
   Resistance to the change was assessed.

Planning phase
The change framework was prepared and communicated, including the reasons for the changes.

Implementation phase
Training of all personnel in the change process. The company dealt with the resistance. Communication was continual. The process was monitored and adjusted.

CHAPTER 12

Converting to a Project-Based Organisation

12. 1 INTRODUCTION

Some organisations, such as construction companies, are totally oriented around projects; others such as hospitals are not, although they do undertake some form of project work from time-to-time.

Some organisations have made the decision to convert from traditional (non-project-based) management to management on a project basis. The packaging of the work is changed. There are various drivers for this change, but generally in the recognition that their work can be seen as a collection of projects (projects are a core business activity) and that project management in a non-project-based organisation is generally difficult to carry out. Functional groupings of people with like expertise and focus, headed by department-centric managers, create strong departmental boundaries inhibiting interaction between departments and project-based work. There is also the belief that a project-based arrangement gives some commercial advantage while simplifying the management process and providing greater control over resources. There may be a realisation that a functional organisation is preventing the achievement of sought after results. Some people suggest that project-based organisations will eventually become the norm, as a way of dealing with the modern business.

The conversion from a non-project-based organisation to a project-based organisation brings with it a number of issues, particularly surrounding people and work culture. Senior management support is needed for the conversion process to occur successfully. In the short term, the change may be disruptive, but it is hoped that this will be repaid in the long term with improved business performance.

12.2 ORGANISATIONAL ISSUES INVOLVED IN CONVERTING TO A PROJECT-BASED ORGANISATION

The issues involved in converting from a non-project-based organisation to a project-based organisation are common to organisational change generally, and include dealing with the 'fear of the unknown', resistance to change, loss of career path, organisational structure issues, changing an organisational culture and procedures, communication of the reason for the change and the status of the change, direct people involvement in the change, and so on.

People

People seem to have an in-built fear of change and, with the amount of change-for-change sake and faddish thinking promoted by management consultants and senior management who have read the latest paperback book on panacea management (Carmichael, 2002, pp. 163-179), seem to have also developed a healthy cynicism for change.

The fear is natural and relates to the uncertainty that the new organisational arrangement brings regarding new roles and relationships and even possible job losses for those who cannot adapt to the new or have the skills to be part of the new. Some people, for example, may have either good technical or management skills and knowledge, but may lack the breadth required to manage projects, and the diversity of activities (both technical and managerial) within a project.

Career paths, and promotional or growth opportunities, that may have been well defined along functional paths, now become unclear, but project management career paths, outside of technical paths, can be established. Functional managers may feel their own power and influence threatened. Some staff recruitment may be necessary.

Conversion to a project-based organisation may alienate some personnel, such that they feel disassociated from the organisation. There is a 'no home syndrome'.

People may have to work with different groups of people. What may have been a comfortable working arrangement, now has to be re-established and groups have to go through their development stages anew. Rivalry may develop between project teams, whereas before the rivalry may have been between functional areas.

With a project-based organisation, there is a dispersal of knowledge and experience around the organisation, compared to a concentration within departments in a non-project-based organisation. At the completion of a project, staff may be relocated and project teams disbanded. The organisation needs to think how learning occurs across the organisation and over time within the organisation. A continuous stream of projects will maintain a project team. With in-house staff, knowledge retention is greater than if work is outsourced.

As projects near completion, there is the question of continuity of employment, and associated fear issues, unless functional or departmental homes are maintained, say within a matrix style arrangement. Will another project exist to which people can be transferred? Will this uncertainty about future work impact on the current project's success?

Structure

There are a number of possible organisational structures that might be adopted – pure task force, matrix and hybrid arrangements. Each represents a different chain of command.

Matrix arrangements represent the least threat, in terms of change, to people because existing functional arrangements and expertise groupings are not lost; this occurs at a possible cost of conflicting loyalties (division of employee allegiance) and dual reporting (to both the departmental manager and to the project manager), because workers also become project team members in a multi-discipline project team, which is dependent on departmental cooperation. Matrix arrangements can be adopted in non-project-based organisations to handle smaller projects, but may prove unsuitable for large projects, a continuing succession of projects, and where strong leadership from the higher organisation levels is lacking.

With a task force arrangement, team members can concentrate solely on the project at hand, but with minimal sharing of resources; the project has priority and other issues are secondary. Team members are physically co-located, multi-disciplined and project oriented. Organisations with task force arrangements might be said to be project-driven. A task force arrangement might be regarded as the more preferential from a project viewpoint.

A hybrid arrangement may, for example, set up projects within each functional area, or any other set up in between pure task force and matrix.

All organisational structures have advantages and disadvantages.

How best to tie the project groups to the parent body, and how best to interface between the groups, needs some consideration. Should projects be given prioritisation? How best to deal with inter-project rivalry and competition; should it be encouraged or tempered? How do project and end-product objectives and constraints flow down from the parent body's objectives and constraints?

Some thought is needed as to what is to be done in-house and what is to be outsourced. This depends on the size and type of projects envisaged and the in-house skills and resources available. Specialist input may be bought in, even on a short-term basis while the transformation to a project basis occurs. Associations with other groups who have project expertise, can assist the learning curve, shorten the transformation time, and assist with any resource shortage.

New administrative and work procedures have to be developed.

Cost

In the original non-project-based arrangement, administrative and clerical support tasks such as procurement may have been handled centrally. On going to a project-based arrangement, such tasks may get decentralised, duplicated and be project driven. Similarly, technology and machinery may be duplicated. There may be less efficient usage of such resources unless there are strong arrangements within the organisation. Decentralisation is not necessarily compulsory within a project scenario however; central influence over resources can be maintained and has advantages because it provides support for all projects but may not be required on a regular basis. As well, hybrid practices may be adopted. All these practices have both 'fors' and 'againsts' in terms of costs.

Coordination

Issues that have to be dealt with by senior management are the overall coordination of projects, the appointment of project managers with appropriate expertise and skills, and the appropriate allocation of resources to projects such that they can progress satisfactorily. A supportive and senior management is called for. A continuity of projects is assumed, either secured externally or generated in-house. Competent coordination and project management attracts further projects.

12.3 PRACTICES FOR DEALING WITH CHANGE ISSUES

In dealing with the change issues, there are a number of recommended practices, including:
- Convey to people the reason for the change.
- Get people's involvement in the change.
- Mobilise commitment. Motivate people.
- Communicate to preempt any misinformation obtained 'on the grapevine', leading to negative attitudes. Perhaps hold regular forums.
- Get senior management support; managers 'walk the talk'.
- Plan the change. Fresh ideas, from outside, may help.
- Properly design the new structure with clear job descriptions.
- Implement any necessary training. Project managers and project team members need appropriate skills; existing staff may be unsuitable in their present form. Future core staff are to have appropriate project skills.
- Monitor and adjust new practices.
- Institutionalise any new approaches.

CHAPTER 13

Technical v Non-Technical Projects

13.1 INTRODUCTION

Much comment is made of technical versus non-technical projects. An example technical project would be one associated with an engineering endeavour. Some mechanical skill and specialist technical knowledge is required. An example non-technical project would be one involving a group of people doing something (non-technical) such as 'moving to a new office', or 'consulting with the community', or 'campaigning for political office', 'feeding the dog' or 'going shopping'. But where the transition from a 'technical' to a 'non-technical' project occurs is unclear.

Historically, project management can trace its origins to industries such as defence, construction, engineering and other technical areas. (So called generic project management is based on such roots.) In more recent years, project management has been embraced almost universally by non-technical people for their work, which they now realise are 'projects'. For example, the employment pages in newspapers now see advertisements seeking to recruit project personnel to inquire into or study various community-related issues, or to report to a government or industry agency. Much has been written and developed relating to the management of technical projects. Such project management is reasonably comprehensive. Non-technical projects lack historical data and experience, and such well defined methods.

The view put forward in this book is one of a generic structure for project management, into which both technical and non-technical projects fit. That is, there is no need to distinguish between the types of projects. There is no advantage in treating non-technical projects differently to technical projects. Certainly they involve different emphases (and some bits disappear completely), but the structure remains the same. This is sometimes expressed as 'a rose is a rose is a rose'; all roses no matter what their type require fundamental necessities of care, but each may also require specific attention due to local soil conditions to successfully bloom. The degree of application and success of methods developed for technical projects depends largely on the type of non-technical project. That is, although a generic structure exists, it is still necessary to distinguish between projects at the detail level.

However, this view is not held by everyone. Some people strenuously insist that the two types of projects must be treated differently. If projects are not differentiated, failure is inevitable.

Because there is no one accepted way of managing projects, disagreement over the treatment of different projects is common.

13.2 DIFFERENCES BETWEEN TECHNICAL AND NON-TECHNICAL PROJECTS

Projects may be categorised according to the degree of definition of their end-product requirements and the degree of definition of the methods used on the project (Fig. 13.1).

Technical projects are predominantly of Type I. Non-technical projects are predominantly of Type IV.

Some specific differences between technical and non-technical projects include:

- Technical projects require people who have specialist technical skills to handle tasks and problems which are part of every technical project. The people work toward the end-product and then may move on to another project. For non-technical projects, most people are candidate project team members.
- Technical projects usually have a well defined specification and scope that has to be satisfied before the project is deemed to be successful. Some non-technical projects have a well defined scope, timeline, budget etc. For example, producing a report to a government or industry agency. However, most non-technical projects are not so well defined. For example, a government inquiry can drag on and continue to consume money due to poor definition and the resulting difficulty in project management.
- There is usually some tangible evidence, a physical end-product (for example, a building, a pipeline, a consumable product), as an outcome of a technical project. This may not be the case in a non-technical project where the end-product may be not easily recognisable. For example, the outcome may be a collection of opinions (community consultation project), or winning/losing by a number of votes (political campaign project).
- The end-product of technical projects is usually measurable in terms of standards (cost, quality, ...), allowing success or failure to be readily identified. The intangible and often ambiguous end-products of non-technical projects are difficult to put in measurable terms, meaning that success or failure is more subjective.
- The ratio of material (and equipment) resources to labour resources is generally different. In a technical project, a lot of capital may be tied up in material resources

Project end-product

		Well defined	Poorly defined
Project methods	Well defined	Type I	Type III
	Poorly defined	Type II	Type IV

Figure 13.1 A classification of project types (source unknown).

compared to labour resources. In a non-technical project, a bigger proportion of capital is directed towards labour resources. Technical projects tend to be material intensive, whereas non-technical projects tend to be labour intensive.

- The boundaries of non-technical projects tend to be less rigid than those of technical projects. It may be difficult to isolate a particular project, as well as to predict the external influences. Technical project boundaries tend to be defined by historical influences.

13.3 MANAGEMENT DIFFERENCES

From a management function point of view, there is no difference between technical and non-technical projects. All require setting objectives and constraints, planning and so on. Also, both project types go through recognisable phases. However there are some detailed management differences:

- Because of the difference in skills required in the teams for technical and non-technical projects, there is consequently a cultural difference between the two team types. Technically-oriented people think and behave differently to non-technically-oriented people. Different people types are said to exercise different portions of their brains. The cultures of the projects are correspondingly different. Managing non-technical people requires a different approach to communication and people skills to that for technically oriented people. Different management styles are called for.

- The range of people suitable as a project manager is less for technical projects. Technical projects require a project manager with a technical background, whereas in a non-technical project, both non-technical and technical people are candidate project managers. In technical projects, some technical expertise is regarded as very desirable and a distinct advantage. Project team members tend to get on well with a project manager with technical knowledge because of the ease in explaining and understanding technical issues. Conversely, team members look down on project managers who do not have technical experience in the area they manage. The communication problems then flow on to give under-performing projects. The choice of project manager is important in the success or failure of a project.

- It is unclear whether a good project manager for a technical project would make a good one for a non-technical project as well. Certainly, such a person could take the initiative in managing the processes. The reverse is clearly untrue, because a project manager, by reason of not having a technical background, is unlikely to have the ability and the confidence to lead a technical team and, most important of all, would be distrusted by her/his subordinates, particularly when the project had to focus on purely technical aspects.

- The role of a project manager generally is open to interpretation, and many consider project management more an art than anything else. Consequently, many people consider themselves suitably qualified to manage any type of project, with or without a grounding in project management tools. There is a tendency in non-technical projects for nearly anyone to consider themselves as suitable project managers.

- Management of non-technical projects could be argued is more difficult because of the unpredictability of human nature. A management style that focuses on people is required, highlighting the need for the correct choice of the project manager for a particular project.

- For non-technical projects, the interpretation of project success may lie in the eye of the beholder. For example, in politics, a political election/re-election campaign may be interpreted as successful irrespective of the way it was run. Much of this is because of the confusion between the project and the often intangible end-product. It would be considered easier to review a technical project to ascertain the reasons why it has failed, and to assess the performance of the project management team.
- With an intangible end-product, the setting of project and end-product objectives, and establishing the project scope becomes difficult to do precisely. With hazy scope, project team morale may be affected. With poorly defined project end requirements and project methods, non-technical projects may not be easy to manage.
- With a heavier emphasis on material resources, technical projects generally have procurement as a major function. Resource management for technical projects involves people, materials and equipment, while resource management for non-technical projects involves mainly, if not solely, people.
- The role of quality is different. Technical projects are normally required to comply with a set of technical specifications; performance is measured against such benchmarks. In non-technical projects, quality is qualitative and non-measurable, and open to interpretation.
- In non-technical projects it may be difficult to establish a critical path. For example, it may be nearly impossible to establish a critical path for a political campaign.
- It may not be straightforward to apply technical project management techniques to non-technical projects. However, non-technical projects can benefit from the application of technical project management techniques.

Case example

Staff redundancy project

This case study highlights the incorrect choice of the project manager in relation to a staff redundancy project. The manager of the project was from a technical background but, in hindsight, the project should have been managed by someone with people skills and experienced in organisational change. The intent of the project was to reduce employee numbers through voluntary redundancy. Throughout the project, staff morale suffered as did productivity. Employees felt abandoned by the company, and the mood within the company changed from one of friendliness and openness to one that was tense and bitter. After the redundancies were complete, employees chose to leave because the company was no longer pleasant to work in. Tact was not used by the project manager, morale was not maintained and employees felt that they did not at any time know where they stood in management's eyes. An alternative project manager could have made the transitional phases for many people easier.

Case example

New corporate identity

Some projects can be a mixture of technical and non-technical components. Consider that of establishing a new corporate identity for a commercial airline. This project involved a number of subprojects which were purely technical, for example the activities related to procuring: new livery for aircraft fleet, new shop signs and decor for ticketing and corporate offices, new stationery, and new uniforms. All these subprojects were technical and could be managed according to established project management practices. However the project also involved non-technical subprojects, for example the activities related to making the new corporate identity known, accepted and recognised, and helping to generate more revenue. This last subproject might be regarded as the most difficult. The reason for the extra difficulty was that the end-product was not a physical quantity and it was difficult to motivate staff towards an uncertain end-product.

Case example

Mergers

Mixed projects with an emphasis on the non-technical side commonly occur in service industries with mergers and take-overs. A typical example was the merger of two banks where the main technical project involved integrating the information systems of the two banks, and the main non-technical project was the integration of two different management systems and two different corporate cultures.

Case example

Airport study

In the community consultation phase of a regional airport study, this subproject within a technical project had an assigned project team, defined tasks, a timeline, a budget in manhour costs and its defined outcome being a section in the study which reported on the findings of the consultation. As the overall study was being carried out by a firm of experienced consultants in the aviation industry, it was relatively straightforward for the non-technical subproject to be couched in accordance with a conventional technical project management format.

Case example

Millennium bug

A project centred on an organisation being ready for the problems that may or may not have been generated by the so-called 'millennium bug'. The project, in some ways, tried to apply technical project management techniques but suffered in a number of areas.

The project was overseen by a steering committee made up of members across the organisation's departments without any one person being designated as the project manager. The committee was to act as the project manager.

The technical representatives from the engineering services department (the department most likely to be impacted by the 'millennium bug') were relatively junior members of staff and were overshadowed by the doom-and-gloom (non-technical) members of the steering group. Whilst the steering group did do some worthwhile things, the project went off the rails with predictions of impending catastrophe and recommendations that all staff leave at the end of the year (prior to the new millennium) be cancelled. Intervention by the organisation's executives was then necessary with corresponding involvement of senior technical staff.

The basic problems seemed to revolve around poor attention to scope and human resources issues. As far as scope was concerned, the project officers in many cases, not being from a technical background, tended to go off on tangents (everything was going to be a problem) and did not appreciate the Pareto rule. As far as human resources were concerned, there was no designated project manager and the wrong people were engaged to carry out the work of the steering group.

Case example

Organisational development

An organisational development project had its genesis in the CEO's attempt to force through another organisational restructure, which was not universally supported. A previous poorly managed restructure had left all involved scarred by the experience. Many simply believed that there was no need for the change. Not to be dissuaded by this lack of support, the CEO instigated a 'Working Relationships Project', an exercise to assess the strengths and weaknesses of the relationships that existed within the organisation and between the organisation and its customers.

This project was typical of many organisational development type projects. It suffered from poor definition and resources, and the length of time needed for the project was difficult to estimate. From the outset a fair degree of cynicism surrounded the intent of the exercise. Some thought that the CEO would use the exercise to push through the need for change, whilst others saw it as an opportunity to allow all members a say in the future structure of the organisation. Those with the latter view were anxious to ensure that the views of all stakeholders were taken into account. This resulted in a significant allocation of staff time in attending meetings, workshops and working groups. There was little realisation, at the start of the exercise, of the necessary staff resources dedicated to the exercise. However, once started, it had to be finished. The project suffered from not having a designated project manager and was managed by a 'politically correct' working party.

About one month after the start of the exercise, the CEO fell into disfavour with the organisation and left. In his absence the exercise continued and the resulting report, that was adopted by organisation, recommended little change and provided a blueprint for future organisational development. A cynic might say that the resultant report was not quite what the CEO had in mind and the organisation might have had a different report had he still been at work.

EXERCISES

1. What is necessary in managing non-technical projects to deal with intangible end-product requirements and the protection of the project boundaries?

2. Would you expect technical people to convert non-technical projects into a form familiar to technical people? If so, is this advantageous or not, in terms of managing a non-technical project?

3. Non-technical people, if left to manage a project under their own devices, seem to approach projects differently to technical people. This may be due to their different culture, education, backgrounds, or way of thinking (brain-part usage). Can you say that this is a better/worse way to go than using an established project management methodology?

4. Which makes the better project manager of a technical project – a person trained in management alone, or a technically trained person with supplementary management skills? Of a non-technical project?

5. Is the culture of a project influenced more by the end-product or by the project personnel?

CHAPTER 14

Projects with Ill-Defined Scope

14.1 SOME ISSUES

Research projects, for example, are notorious for their lack of scope definition. Much of the work is 'follow your own nose' or 'go in a direction which looks fruitful' or even serendipity (fortunate discovery by accident). The amount of work, the cost of the work, and the time to do this are very open-ended. Research projects are a challenge for anyone used to projects with well defined scope.

As such, what procedures should be adopted in managing projects such as research projects, where the scope is never really defined until the project ends? A related question is how important is it having scope well defined in projects?

The poor definition of scope arises not so much from the project objectives and constraints, which can be defined usually fairly well, but rather from the unclear way of getting to the end-product and uncertain end-product.

Cost plus (prime cost) contracts are a common way of procuring work which is ill-defined beforehand. Bonuses/penalties can be added if targets related to, for example, cost, time and quality can be agreed. Generally such contracts require more administration input from the owner overseeing the contractor. (Carmichael, 2000)

14.2 SCOPE DEFINITION

Scope is what is involved in undertaking the project, the extent of the work to be undertaken, what work is contained in the project (perhaps by defining what work is not included in the project, in effect the boundaries to the project) that leads to the end-product. Scope is fully described by listing all the project activities.

There is much confusion over the usage of the term scope. Some writers wrongly include the end-product, resources, quality standards, performance measurement, acceptance criteria, assumptions, approach or methodology to be used, justification, approving body, cost implications of options, timing, environmental (natural) impact, employment considerations, political ramifications etc in the definition. While useful information to have, it is not scope, and should be clearly distinguished from scope. Some see a scope statement as a description of deliverables (end-product plus state – See Chapter 16 – final (terminal) conditions for the project) or 'what the customer expects to see as a result of the project'. The process and the result are confused. It appears that the word scope is used by most people in the sense of Alice in Wonderland:

'When I use a word,' Humpty Dumpty said in a rather scornful tone, 'it means just what I choose it to mean, neither more nor less'.
'The question is', said Alice, 'whether you can make words mean so many different things'.

('Through the Looking Glass', Ch. 6, Lewis Carroll)

The biggest transgression though is the sloppy, interchangeable use of the terms 'objective', 'constraint' and 'scope'; few people appreciate the distinction and why there is a distinction.

Commonly the carriage is put before the horse, such that the scope determines the objectives and the constraints. This confusion and imprecision stifles the understanding of project management.

14.3   GENERAL

Most people would agree that it is important, even essential, to define the project scope well, in order to ensure that the team knows what has to be done, the project has direction, the boundaries of the project are established, the project delivers the appropriate end-product at the appropriate time etc. Most people would also agree that it is important to have this scope defined as early as possible in a project, because all subsequent project work, such as organising, developing budgets, schedules, monitoring and control, depends on it. Changes made later in a project, rather than earlier, may be more costly. An inadequately defined scope and frequent scope changes are seen as main reasons for project under-performance or a perception of under-performance.

Some people argue that scope issues are the most important issues facing a project manager. A project may be constantly under pressure to accommodate changes and modifications.

A project, with inadequate scope definition, is a challenge for any project manager. The scope may develop iteratively with the development of the project. Gauging a project's success becomes difficult, and indicators of success may be developed iteratively. The end result may be viewed by an observer as either very positive or an absolute waste of time. The end result may not be possible to achieve. The project may be regarded as a technical success but a commercial failure.

There is no clear direction; uncertainty is present. Standards are necessarily set as prescriptive. There is every chance that the project may drift away from its original intent, an initial focus is lost, factors not initially considered may be found to significantly influence the outcome, and constraints of time, cost and performance will not be met or even go close to being met. A decision as to when the project is to be terminated may be open. People lose sight of what they are trying to achieve. Agreement may be lost between the project team, the owner, project manager and other parties.

An approach, as systematic as possible, yet necessarily different to that for projects with well defined scope, is needed, though the principles remain the same between the two project types. There may be, for example, a limited budget to undertake work whose scope is unknown or open because of the nature of the work being performed. Science-based research projects are commonly of this form; the scope appears unbounded, yet there are usually funding constraints; the researcher works until the money runs out, irrespective of whether a satisfactory outcome to the research has been reached. The intent of research projects is to extend the boundaries of current knowledge, and the path is either difficult or impossible to define. The project descriptions are necessarily loose so as to not constrain the formulation of novel ideas and solutions to any problems that may arise.

Projects, which initially have a poorly defined scope, but as they progress, the scope becomes defined and allows the project to be managed satisfactorily, are not addressed here. Most projects, early on, may have poor scope definition, but here discussion is restricted to projects where it is difficult at any stage to define the scope. Also, projects which are capable of having a well defined scope but which don't, are also excluded from discussion. Such projects are common because many people are 'doers', and like to get involved in the solution or challenge before defining the problem. This situation is a reflection of poor planning, and frequently leads to crisis management.

14.4 PROCEDURES FOR MANAGING PROJECTS WITH ILL-DEFINED SCOPE

There are some procedures that can be put in place to manage projects with ill-defined scope, or projects in uncertain and/or rapidly changing environments.

Compensating detail

The lack of detail in the project scope is balanced by carrying out other elements of the project in more detail than would normally be required. Other management aspects take on greater importance, and give an increased degree of management of the evolution of a project than would otherwise be necessary in a well defined project.

Feasibility

A feasibility study carried out before the project proper may help establish the viability of the project, before significant money is invested. The study would help highlight the risks (technical and commercial) involved.

Approval

Project feasibility is always uncertain, and hence gaining approval to proceed with an ill-defined project is always uncertain. Since it is not possible to establish exactly what has to be done, it is not possible to say how it is to be done and how much it will cost. Approval instead might be expressed in terms of ceilings on time, money and resources, that is the upper limits that the project owner is prepared to support. However this does not ensure an end-product, or a desired end-product, even if all ceilings are reached. The ceilings may be reached, no end-product achieved and the project abandoned.

On-going approval and continuous review may be carried out in a step-by-step approach as the project progresses. The step may be a time interval (regular or irregular), or a certain expenditure. In effect, sub-ceilings are set throughout the project, and as each sub-ceiling is reached, the project is reviewed, and approval is or is not given to proceed.

The authorisation and funding for the following step is limited by the sub-ceilings set for that step. At each step, the project owner can assess time/cost-so-far and estimated time/cost-to-go against the potential of reaching the desired end-product, past and future obstacles, and a decision can be made to continue or not. On-going approval avoids having to commit to an undefined effort for an uncertain final result up-front. An end-product is still not guaranteed however.

Objectives and constraints

Addressing or understanding the project's and end-product's objectives and constraints (time frame, resource expenditure, ...) at the project outset gives the project some direction, focus (a light at the end of the tunnel) and boundaries.

A clear understanding of the intent of the project, and a focus on that intent is maintained and continually reiterated throughout the project. The nature of the project work may be unknown and the path to the end-product may be unknown for much of the project, but the intent of the project is always made clear. With continual project review, there may be a change in intent during the project's life.

The objectives and constraints may be set for not only the overall project but for each part of the project.

Scope delineation

Without a broad initial scope, scope delineation (primarily through the device of a work breakdown structure) is not possible. Without scope delineation, budgets and schedules are not possible, and following that project control is not possible. Scope changes become blurred with normal project development. The project is uncontrollable; the final cost and time are unknown.

However, it might be possible to build up a picture of the scope by assembling blocks of work (subprojects, sub-subprojects, ..., activities/tasks), in the reverse direction to a work breakdown structure. These blocks of work represent what people see as being necessary work. The scope of some of the subprojects may be definable, while the scope of the project may not be. For those subprojects whose scope is not definable, the recognition of this can be highlighted. Previous similar projects, generic checklists or standard operating procedures may assist in firming the scope of some of these subprojects. Management at the subproject level may be easier than at the project level.

A clearer picture of the scope may evolve as the project progresses. As each subproject is completed, the project direction and next subprojects are determined. Knowledge about the project develops as the project proceeds.

Planning

Flexible schedules and budgets are prepared allowing for uncertainties, and are indicative of how the project will proceed. Unanticipated problems will cause deviations from the initial plan. Schedules include float and contingencies (time reserves), and budgets include contingencies (cost reserves). Contingency planning is carried out.

Control

Some form of control, to appease the financial backers of a project, can be obtained if a step-by-step approach to authorisation is used. Project milestones/freeze points are identified and authorisation only given to proceed as each milestone is completed. The milestones may only become known as the project progresses. Information on time, cost-to-date, manhours-to-date etc may be continually collected and a decision is made on the future direction of the project and the re-allocation of resources. The direction in which the project is heading may undergo change if progress is not liked.

If applied at the smallest activity level, control may be easier, and the uncertainty may be less. Each completed activity might be regarded as a milestone. This higher degree of project control and higher frequency of milestone analysis, may mean more time is spent in reviews, however.

Some further form of control can be obtained if ongoing information that is collected is compared with that from similar past projects. An assessment may then be made as to whether progress is satisfactory.

Project plans or a project baseline are updated continually – progressive baseline definition. The documents become dynamic. Although this is considered good practice for any project, it takes on added importance for projects whose scope is evolving.

In parallel, the scope may be becoming better defined and this may be utilised as it comes to hand. For example, in a medical research project, the initial drug discovery phase may be very indefinite, but the following clinical trials may be able to have the scope well defined based on previous practices.

Good management is still necessary to ensure that the work is being done efficiently, and unnecessary resources aren't being used. All changes are recorded, rebudgeted and approved. Progress and expenditure are tracked.

In projects requiring creativity, for example research projects, imposing controls may inhibit this creativity and the sharing of knowledge across the team. Controls are required to be flexible enough to account for new factors that may arise, to be general enough to accommodate changes, and to assist with maintaining project focus.

Iterative end-product development

In product and software development projects, iterations may be used in an attempt to establish what is the end-product that is wanted by the owner. This in turn leads to iterations on the project scope. An initial product is developed based on what the owner believes s/he wants. The product is evaluated by the owner and then modified based on this evaluation. Several iterations may occur before a final product is settled upon.

The process can be open-ended unless some way of ending the iterations, namely satisfying the owner's requirements, is given. Gold plating and the adding of marginally useful features may occur if the iterations go on for too long.

Each iteration can be timed and costed, and hence controlled. The more definite the owner's requirements are stated up-front, the fewer the iterations that will be necessary.

Loss of focus

Projects with ill-defined scope can lose their focus. Something more interesting catches attention. A line of lesser resistance is followed. Sideline issues take prominence over the main (Pareto) issues. Work is unnecessarily undertaken to too fine a level of detail in areas where it can be, at the expense of depth in other areas.

Attention needs to remain focused, and all project matters put in perspective.

Team member roles

The individual responsibilities of the project team members, including the project manager, are defined in detail. A clear chain of command is in place. Everyone knows who is responsibility for any project outcome. Team members are given responsibilities for tasks.

Communication

Communication between project stakeholders is important in the success of any project, but perhaps more so when the scope is poorly defined. Poor communication results in resource wastage and inefficiency. Information for the right person in a timely manner is needed. In research projects, team members tend to be highly motivated and interested in each other's work.

Management style

There may be a need for the project manager to have a high level of people skills and an ability to guide the efforts of team members rather than have commercial skills such as those used in contract dispute resolution.

Case example

Operations

The project involved large capital expenditure. The scope document was prepared by operations managers who did not have the expertise in preparing such a document, and was developed based on input from the contract project manager. They created a document that appeared to them to define a satisfactory scope for the project. Early on in the project, they found that what the contract project manager had given them was a lot of unnecessary things and an expenditure that was considerably over budget. The reason for this was that the scope defined what was required in broad terms, but did not specify exactly what needed to be done. As a consequence, the contract project manager had included all the 'nice to haves' because she was not responsible for the budget. The contract project manager was being paid on a reimbursible basis up to a limit, but the project was not complete by the time that limit had been reached. The contract project manager hoped to make as much money out of the job as she could and so without a clear scope and good control, she was able to do more than necessary.

Case example

Research projects

Research projects use scientific procedures to inquire into the unknown. Time and cost elements of the project are difficult to estimate. The scope of a research project is not known until the project is completed

Team members need freedom to fully utilise their creative skills and expertise, whilst there still needs to be time, cost and direction control exerted over the project.

Case example

Aquatic centre construction

The construction of an aquatic centre was a project where ill-defined scope led to significant delays and cost overruns. The design was documented and construction went to tender. During the tender period, the tenderers were extremely critical of the documentation and commented that the design, as documented, could not be built. Two tenders were received (for about twice the initial estimate), and both were subject to numerous qualifications.

The design consultant failed to advise the owner of the impact owner-requested changes would have on project cost. Being from out of town, the designer also failed to appreciate particular local characteristics. The owner's representative had no project management experience and had total trust in the consultant. Needless to say the owner was shocked at the cost blow out, sacked the consultant, and the owner's representative was promoted sideways and was forbidden to dabble in any future projects.

The project was resurrected a year later under different people. The project team for the resurrected project was recruited around a detailed project brief that had a detailed project scope related to the cost and site specific constraints, and project management that included time, cost and quality elements. The centre was then delivered.

Case example

Health research

The procedures adopted by one health research group include:
- Research grant proposals include details of what the project proposes to do, what methods are to be used, possible alternative methods if the original methods prove inadequate, and the proposed length of the project.
- The total project time is then divided into increments and mini-milestones set.
- Regular project meetings are held. Results and problems are discussed and the plan of attack until the next meeting decided upon. Advice is sort from people outside the project – these are often able to see problems and offer solutions that people involved with the project are unable to see or don't consider.
- Weekly laboratory talks are held in which people from different groups address all the laboratory staff and discuss what they've done since they last spoke, any results they have obtained and any problems they have had. Each group speaks in weekly rotation. People from other groups can then offer advice – in many cases they have encountered similar problems in the past but because they are involved in different projects they are not aware that someone else is having that difficulty. Laboratory talks ensure everyone has a reasonably good idea of what is happening throughout the laboratory.
- Methods are only tried for a set period of time, or for a certain number of times before they are discarded and alternative methods tried.
- Publications are constantly scanned for new methods, or methods that can be altered to suit a project.
- Problems are not allowed to drag on. They are instead discussed with supervisors as they happen and, hopefully, solved in the shortest possible time.
- Groups collaborate with related project groups in other establishments, providing a wider knowledge and experience base. Most research staff also have a network of previous workmates and acquaintances that they can call upon for help and advice.
- Adherence to work deadlines very much depends upon the individual, but as most people are employed on twelve-month contracts, there is a reasonably high incentive to maintain good work practices. The awarding of grant money also depends upon past research performance.

Case example

Livestock

A project was initiated by a livestock corporation which engaged a consultant to explore the possible use of automated slaughter technology (AST). It ended in a failed attempt to commission a commercial automated slaughter facility many years after the project was commenced. The project was regarded as a technical success but a commercial failure.

The project was broken up into a large number of finite task blocks. Each task was technically simple and was carried out with a clear understanding of the overall objective, culminating in a milestone, thereby reducing the effects of any uncertainties. At a milestone point, the project was reviewed in terms of the larger picture and adjustments were made as necessary.

This approach enabled a high degree of project control, although during the life of the AST project it did somewhat resemble an exceedingly long review meeting, as the nature of the project led to a very high frequency of milestones and associated analysis. Yet this approach did provide a basis for the evaluation of progress by identifying the critical elements of the project and expressing them in terms of time and effort. Any deviation from the plan was regarded as a signal for action which resulted in a re-allocation of resources to recover the time scale originally planned or a reformulation of the plan to reflect the new realities of the project.

It was felt that the team members carrying out this project were highly motivated and well equipped with the appropriate tools – planning, group decision making, problem solving and communication – and that if control was imposed on them, the sharing of skills and project insight together with the creative synergy would fundamentally disappear.

Each team member, as a concession, was required to give a weekly financial update on the task to which they were assigned, and this was presented as a combined weekly report to the livestock corporation.

It was also found that with the risk perceptions being formally shared amongst the critical team members a greater degree of project team cohesion existed.

Case examples

Mining

(a) An example research/problem solving project was the construction and operation of a pilot plant examining material consolidation and thickeners for the management of tailings. This was an extremely high profile issue within the organisation and required immediate results because it involved a newly operational mine. Ideas were generated by design staff on how a test rig should be built and operated. The rigs then had to be built from what materials were on hand, to perform the function as required. The operation and testing was overseen by the research side of the project team. The overall project consisted of five separate subprojects which operated independently of each other, but each forming a step in the overall project.

(b) In a project involving the testing of different vendor products, focus was lost on 'why it isn't working' rather than 'how to get it to work'. The research quickly became focused on sideline issues and the project intent was lost sight of.

Research had found that the high iron content of the water at the site affected the results of the vendors' products rendering them unsuitable. The research people then set about detailed testing of the water and looking at ways to change the quality of the site water supply to suit the vendor's product rather than finding products to suit the water or looking at the exact needs of better quality water. Once the issue of the water had been identified, discussions were held with all the people involved and their needs for good quality water were established. This resulted in small amounts of filtered water being required rather than the entire site water supply, with ongoing monitoring of the main water source for abnormalities.

(c) Problems in the level of detail became an issue with a testing of thickeners program. One of the subprojects involved the design and construction of a manually operated conventional thickener, on an extremely tight budget in a tight time frame, to gain indicative results which might be further pursued. As the construction neared completion and operation commenced, the research team decided that there was not enough information being gathered and that the operation should become automated. This resulted in pressure gauges, electronic flow meters, multiple sampling points, computer interfaces and controls being installed. Time and cost were of little importance because the information was needed in the name of science. Part of this problem extended from another research project which was being run concurrently utilising similar equipment at the site. As the opportunity arose they thought they would extend their testing program from one rig to two. Whilst this may have been an opportune moment for the other project, it held up this project and pushed out the cost and time of the project.

(d) Much of a project had been completed, however this came after five months of an initial six week testing period. Testing continued on a remaining part. The costs involved with changing the process operation were considered with the assumption that this final part would be successful. From this cost estimate it was considered that the operation would not be financially viable, and as such the testing was halted. This did meet with some disappointment, because the team was close to successfully developing a new method of deposition. It required the management review process to highlight that, although this part of the project was close to realisation, the greater project was not viable and a halt to the project was called.

CHAPTER 15

Fast-Tracked Projects

15.1 INTRODUCTION

A project is said to be fast-tracked if its phases overlap. (Fig. 15.1). A phase is started before the previous phase is complete. Project phases are run in parallel where they may normally be run sequentially.

At the activity level, a similar situation occurs when start-to-start or finish-to-finish links are specified on the activity network, in place of more usual finish-to-start links. This has the effect of completing the work sooner, which is the intent of fast-tracking.

A common occurrence of the fast-tracked approach is where the design and construction phases overlap; that is, the design is incomplete before construction starts. See for example, Figure 15.2

The primary benefit of fast-tracking is a reduced schedule, quicker completion time for the project, and hence earlier delivery of the end-product. The actual project cost could be expected to increase, in return for an earlier return on money invested, for example earlier

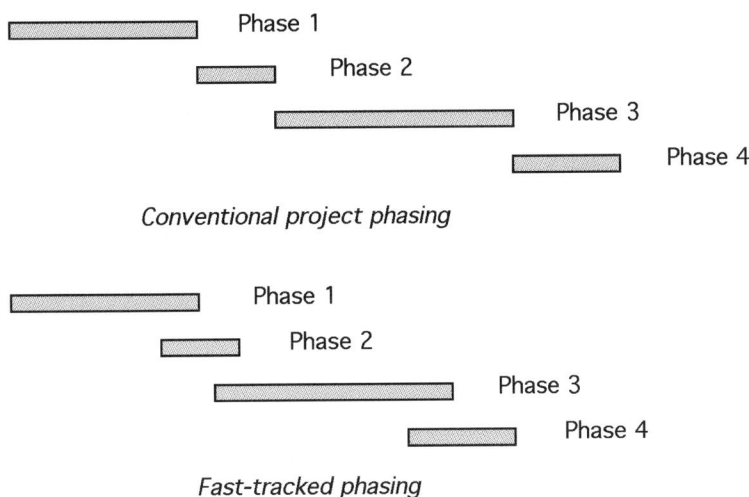

Phase 1

Phase 2

Phase 3

Phase 4

Conventional project phasing

Phase 1

Phase 2

Phase 3

Phase 4

Fast-tracked phasing

Figure 15.1 Bar chart showing comparison of conventional and fast-tracked approaches.

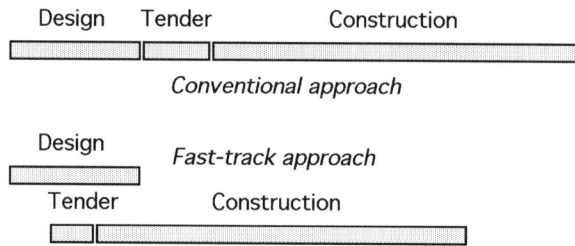

Figure 15.2 Bar chart showing overlap of design and construction phases.

collection of rental, earlier sale, and decreased cost of borrowed money. Related bene-fits of fast-tracking may include, for example, an early market monopoly, a brand-name advantage over competitors, and a good image and reputation for delivering ahead of time. There may be some political commitment driving the need to fast-track a project; political mileage may be scored by completing a project by an announced particular date.

The downside with fast-tracking is the associated peculiar managerial problems. These may not be an issue if the project work is routine and well understood, but in other cases these problems could manifest themselves in, for example, deadlines not being met, extra work, rework and increased costs and resources. Good management practices are needed to reign in any potential troubles. The greater the overlap between phases, the greater could be expected the managerial problems; a decision is needed as to what is the appro-priate amount of overlap.

There may be economy in time, but not necessarily in money. A cheaper alternative, for example, may be traded-off against timeliness.

Straightforward projects are regarded as better suited to fast-tracking. A complicated project adds further issues to those associated with fast-tracking.

Comparisons given below are relative to conventional, that is non fast-tracked, pro-jects. Actual experiments cannot be carried out comparing non fast-tracked and fast-trac-ked projects because of the unique nature of projects and the multiple variables present in a project that influence a project's outcome. Interestingly this hasn't stopped many writers from doing comparisons and drawing conclusions. However, such conclusions should be viewed with scepticism.

15.2 MANAGERIAL PROBLEMS PECULIAR TO FAST-TRACKED PROJECTS

There are a number of managerial problems peculiar to fast-tracked projects.

Approvals

Fast-tracking cannot tolerate any delays in any required approvals, else the tight program will not be met. Staged approvals may be a preferred way to go, but this introduces its own problems.

Owner input

Owners could expect less input to the project, and hence possibly less satisfaction. Less time is available to consider alternatives. With a lay owner, it may be difficult to explain what is happening, given the incomplete documentation and speed of project progress. In passing, the owner's requirements may be overlooked. Special communication with the owner may be required.

Planning

In order to be able to respond to changing conditions as the project progresses, some flexibility in planning is needed. Planning, as for regular projects is carried out, but in addition could be expected to evolve more as the project progresses. With more constraints applying to project activities, there is less free float on activities making the management of the project more difficult.

Lack of documentation, and communication problems

Because of the speed at which the project is being forced, there could be expected to be a shortage of documentation, as well as incomplete communication, as input to decision making. Planning may be that bit more difficult as well. In the absence of required information, there is a need to make assumptions. Confusion amongst team members may also result, as could errors, omissions or rework be expected. Verbal communication may become the dominant form of communication.

Change and rework

There is a greater chance of change and rework (including redesign) on a fast-tracked project, as assumptions made turn out to be not completely correct. For example, the design may be revised, although construction may be complete based on the original design. As the project progresses, the chance of change, demolition and rework could be expected to increase, as current work depends on a growing volume of previous work and assumptions. This adds greater pressures to a perhaps already pressured project team as the project progresses.

Rework, in turn, leads to schedule delays, loss of productivity and increased costs.

Cost control

Fast-tracking could be expected to lead to increased costs because of the conservatism of assumptions in the earlier phases, inefficiencies in design, increased managerial problems, rework etc. There is also less control over project costs.

Disputes

Many changes leading to the need to modify or demolish completed work may lead to many claims from the contractor for extra costs and time. With speed of progress the driving force behind fast-tracked projects, there is a need to resolve disputes and problems quickly as they arise, if the benefits of fast-tracking are to be realised. This essentially excludes the use of litigation and conventional arbitration, and points toward some of the alternative dispute resolution mechanisms such as disputes review boards.

With the shortage of documentation, incomplete communication, change and rework, more disputes could be expected compared to regular projects.

Scope

It could be expected that it would be more expensive or more difficult for an owner to make changes to the scope of a project if it is being fast-tracked because further changes would add to the downstream complications already there because of fast-tracking.

Quality

With the emphasis on speed, decisions are made hurriedly, and there can be a temptation to reduce quality. Fast-tracked projects may not be delivered to the standards expected by the owner.

Pressure to keep costs down, while still accelerating the work, can result in a reduction in quality.

Procurement

Hasty selection and installation requirements may lead to incorrect materials/product selection or incorrect quantities. Tender periods for all matters are shorter, leading to increased risks for the tenderers. An allowance for these increased risks may be included in the tender prices.

The main contractor's tender costs are higher; for example in design-and-construct (D&C) delivery, preliminary design and extra investigations need to be carried out by the contractor before submitting a proposal.

Material requirements, and the procurement of these, for the later phases of a fast-tracked project may be uncertain.

The owner's preferred method of tendering and preferred delivery method may not be compatible with fast-tracking. The owner, in such a case is forced into unfamiliar territory.

Coordination

With work being conducted in parallel, there is increased pressure placed on the coordination of human resources, consultants, subcontractors, suppliers, equipment and activities.

With tight time constraints, there is pressure on people to perform. Any delay by one person or one group flows on to other people and groups.

15.3 PRACTICES FOR DEALING WITH THESE PROBLEMS

Assumptions

With lack of information, assumptions need to be made. Necessarily these are conservative (and more costly) – stronger, bigger, ... in order to not under-anticipate future requirements, to not restrict future end-product possibilities, and to prevent rework and delays (and associated loss of staff morale). For example, assumptions on footings for a building need to anticipate all possible future superstructures.

Procurement

Lead times of items (materials and products) are identified and orders are placed to ensure they do not delay the project. While this is so in conventional projects, it takes on added importance in fast-tracked projects. This is in a project scenario where specifications and design could be expected to be incomplete.

Procurement practices may expedite the tender process, and engage a preferred vendor on the basis that the material or product specification is yet to come, rather than beginning the tender process only after the specification is complete. Preferred vendors may be assembled, and purchase orders issued to these vendors when the specification or design becomes fully known.

Buyer-seller relationships, preferably long-term, are established. Partnering-style thinking could be adopted.

As much detail as possible is included in the tender documents.

Scope

The scope is clearly delineated and everyone is made aware of it. The process of fast-tracking a project appears to be most suitable for projects with well defined scope. Scope changes result in additional costs.

The preparation of a comprehensive design brief assists in minimising design changes.

Risk

Fast-tracked projects could be expected to introduce more risk events/sources. These are identified, and the associated risks are managed to an acceptable level. There could also be expected to be less detection time available to rectify problems.

Issues may arise that have not been previously considered. For example, the design assumptions may prove to be incorrect. Risk management under such circumstances is debatable.

Cost control

It may be necessary to commit extra funds as the project progresses in order to achieve a pre-assigned completion date. Delivery methods and contract payment types can be chosen such that the owner's exposure to additional costs is small. (Carmichael, 2000)

Planning

A structured and systematic planning process enables the identification of possible over-lapping of activities, the identification of problems, and the appropriate allocation of time to activities, and permits the planning to be dynamic. The value of planning, particularly early in the project, should be recognised.

Project crashing or compression within the critical path method provides the ability to shorten a project's duration without changing the extent of work. A cost-time trade-off may be possible giving project compression for least incremental cost. Any savings by accelerating one part of the program may not be realised as there may still be some other time-controlling component.

Preliminary concepts, plans and budgets are made more definite as information comes to hand.

Changes or delays are tracked. Replanning occurs, all relevant parties are notified and resource requirements are adjusted.

Working relationships

Long-term working relationships, such as those promoted through partnering, are encou-raged, in order to create trust and a team spirit. Additionally, time usually set aside for tendering is eliminated.

Project team

Having the right people with the right skills, while important for conventional projects, takes on added significance with the nature of fast-tracked projects. Project manager selection is important. Team members are empowered and given trust. Full and proper delegation is needed. Good communication channels are set up between team members. Team members fully cooperate with each other.

The concept of the project team is enlarged to include most stakeholders, such as sup-pliers, downstream operators, and the owner. All members of this greater team contribute to decisions affecting them and the direction of the project that affects them.

Additional management effort is required for fast-tracked projects. The project mana-ger needs to devote extra time to the job.

Disputes

Dispute avoidance practices are adopted wherever possible (Carmichael, 2002). For example, the contractor may adopt an open book approach with regard to schedules and budgets. However with the recognition that not all disputes are avoidable, something like a disputes review board may be introduced.

Incentives

Incentives promote a willingness to meet or better schedules. Incentives may be introduced for all operations including the supply of both goods and services. Disincentives
for running late (for example a bonus foregone for running over time) may also work but
would be less preferred to that of incentives.

Information

Ensuring all project participants have access to timely and up-to-date project information
(data, plans, designs, ...) facilitates a smoothly run project. Good communication, as in
any project, is important. For example, anything constructed before the design is complete
needs communication at the time, of an as-built diagram to the designers for compliance
checking.

Trends

Key project indicators (cost, schedule, resources, delays, variations, ...) are monitored and
reported in a timely fashion. Trends are noted, and management action taken in order that
the project does not deviate too far from original intentions. Flow-on effects of delays are
tracked. What is measured, is managed.

Case examples

Gold mine

(a) The development of a gold mine's concentrator was a fast-tracked project. Major
parts, such as tank cells for the gold flotation circuit, were ordered well in advance
due to the long lead times required. The cells were originally designed as 100 m^3, but
ended up as 150 m^3. However, the footings, buildings and cranes had already been
constructed and installed for the smaller cells. This created considerable problems
– there was insufficient space to accommodate the new cells as per the new design. A
solution was however, found. The 150 m^3 cells, usually aligned in rows, were stag-
gered to fit within the designed space. While this caused additional problems with
the construction of launders and piping, there was a capital cost benefit to outweigh
the time lost on redesign. If the original cells had been installed as designed, 20 were
required. The larger cells only required that 15 be installed.

(b) In the design definition phase of a gold mine's high grade concentrator project,
the wrong footprint size was used for the building. Significant time and money was
expended in expanding the building. The project was progressing at such a speed
that the detail was missed – it should have been picked up much earlier. Ensuring
that key staff had access to the relevant information could potentially have saved
considerable time and money.

(c) An extension for a gold mine's main office was ordered before final design was completed, although agreement on the basic design had been reached. Work was tendered and the building work awarded to a contractor. Following this, there was a decision to make some changes within the existing office, which impacted on the design of the new extension. It was found that the changes, although minor, would require moving supports. The building construction was temporarily halted; the contractor moved on to new projects and the mine lost its position in the queue. As a result, the building was delayed by months. However some staff still had to move, and so were forced to move twice. Better planning and communication would have enabled earlier detection of the changes required. Eventually the cost over-run resulting from the delays and changed specifications halted the development of the extension. The changes to the main office that had triggered the change of the extension design did not eventuate.

Case example

Apartment building

A fast-tracked apartment building resulted in high levels of end-user dissatisfaction. To save time, the developer chose to build to a minimum building code specification. The minimal inter-apartment walls resulted in a lack of privacy between apartments. This resulted in threats of litigation from the residents. The developer was then involved in expensive retrofitting of the building to satisfy privacy requirements.

Case example

Power cable

Poor project management skills resulted in a fast-tracked project being halted completely. A power organisation lay many kilometres of high voltage cable to connect a region to the existing electricity market. Unfortunately the organisation's proposed route raised health concerns in the local communities that it went near. No alternative route was considered, due to cost considerations,. The project work was delayed for several weeks outside one property, where a standoff developed between a contractor and a property owner. The organisation was further criticised and faced litigation for environmental (natural) damage in environmentally (natural) sensitive areas. Poor site management resulted in contractors over-clearing environmentally (natural) sensitive areas. The local government ordered work on the project to cease until proper public consultation had been conducted.

Case example

Gas piping

Following a fire at a gas plant, a company fast-tracked the laying of gas pipes to the plant. The scheduled time frame of eighteen months, was reduced to six months. To do this involved accelerated negotiations with native landowners and farmers, the procurement of pipes from overseas sources, and labour working 12-hour days and 7-day weeks. Risk assessment determined inclement weather and terrain as the factors likely to delay the project.

Case example

Busway

A transit project involved constructing a new roadway system to be used exclusively by buses.

In mid year 1, the owner announced that the busway would be operating in time for a major event in late year 3. This provided a very short lead time resulting in the project being fast-tracked. This was achieved by using a design-and-construct (D&C) contract with a contract period of one and a half years. The D&C contract also enabled the owner to seek an innovative design.

The contract was awarded in late year 1. Detailed design commenced almost immediately and was scheduled to be completed in mid year 2. The design team struggled to meet the program for several reasons, including complexity of the project, owner-initiated changes to the scope, delays in design approvals and lack of resources. Eventually the design was completed in early year 3.

Construction commenced in early year 2 when 20% of the design was completed. The contractor divided the work into several defined areas and treated them as separate projects. The construction program was also fast-tracked where possible. The project was disrupted after construction commenced when the busway was enlarged so as to cater for light rail (trams) at a future date. This created design and programming problems for the contractor and led to substantial variations but the end date could not be extended.

Despite the problems listed above, the main completion dates and quality standards were achieved.

Case examples

Bridges and highways

(a) A project involved a bridge replacement and the realignment of several kilometres of a highway.

The project was fast-tracked as a result of another project being delayed due to environmental (natural) issues and, as a consequence, this project was elevated on the priority list and certain funding allocations had to be achieved.

Fortunately, several preliminary studies were undertaken for the project in earlier years. However, the project had not been able to proceed further because of funding arrangements.

Essentially, the project was fast-tracked by overlapping several pre-construction phases by using in-house owner resources and external consultants. The earlier studies proved to be most valuable and were generally updated to satisfy current requirements.

(b) A project consisted of two separate highway sections. The two sections were upgraded as part of an overall strategy to improve the highway.

One section went through a mountainous range and involved a major cutting. The second section passed through a township and involved the replacement of a bridge and road realignment.

Funding for each section was from different sources. Due to changes to funding arrangements, the priority for this project was increased and it was necessary to fast-track the program to meet the owner's revised time frame.

A consulting firm was awarded a contract to undertake preliminary geotechnical and natural environment assessments and to prepare a concept design. After advice that the project had to be fast-tracked, the scope of the original contract was increased and variations were issued to the consulting firm to undertake a final geotechnical assessment, a review of environmental (natural) factors and the detailed design. This enabled several months to be deleted from the program due to the time involved in the preparation of tenders, required approvals and the tender process. The owner's time frame was achieved.

The overlapping of several linked phases of the project was required in order to achieve the owner's requirements. For example, the concept design had not been finalised before the detailed design commenced. Also, the findings of the review of environmental (natural) factors had to be incorporated into the detailed design, although the two phases were being undertaken concurrently.

Such overlapping created obvious risks and the project could not have had a successful outcome unless the parties involved had confidence in each other and worked in a cooperative manner. Also, the major changes to the scope of the original contract had to be in areas in which the consultant has expertise.

Case example

Dockyard

The project discussed here is the development of a new dockyard.

During the construction of the dockyard's wedge car slipway, the project was fast-tracked and often construction proceeded ahead of the design. An engineer was employed by the builder to track the differences between the design and what was being built and report these to the designer. He was also responsible for tracking and monitoring the designer's progress with the design. On a daily basis he would inspect the construction of the work, report changes, site conditions and any modifications that may have been required to comply with the design.

Many structural items had to await drawings, but most of the civil work was completed without design. As the owner was also the builder, the owner's engineer completed the civil design in house without the need to produce expensive drawings.

A program was produced to complete the work by a nominated date. The program was tracked, on a weekly basis, by the project manager with his key construction personnel. However the schedule slid by two months due to design changes implemented by the owner. Although the dockyard became operational, some of the facility had not been completed and continued to be constructed.

The owner rarely reviewed the drawings and often implemented changes in the field; the original scope of the project was changed many times. This made it difficult to track the design changes. The owner was often unsatisfied with the sizing and apparent uneconomical use of construction materials. A change was implemented, after the design was started, to go from a maximum 3000 t ship at a maximum length of 100 m to a 6000 t ship at a maximum length of 130 m. This significantly delayed the completion of the design, because the member capacities had to be recalculated for the new loading. As a result, additional members were required to be added to the existing structure.

The wedge car was completed and the final design drawings arrived afterwards requiring only a few minor changes. Construction of the berthing dolphins awaited the detailed design drawings; many changes were made on site and the construction proceeded from preliminary design drawings.

An overseas-based consultancy was employed to provide the structural design for the construction of the wedge car, transfer car and berthing dolphins. The construction proceeded rapidly, much faster than the design, and due to poor planning, the designer was constantly under pressure to finalise the drawings. If the project had been planned from the owner's scope (that was never produced) and tracked accordingly, a lot of unnecessary pressure could have been avoided.

Due to time constraints in the design of other components, a full geotechnical investigation was not conducted for the design of the rock anchors, for the anchor block for the main winch cables. As a result, a greater size and number of anchors were used in the construction of the anchor block.

As the construction proceeded ahead of the design, in many cases the construction team continued without shop drawings and without the finalised details, especially with respect to the connections of members. In most cases, experienced tradespeople used their knowledge of the relevant standards to continue working. These details were later checked on site by the designer. This was not without mishap, as in one case an additional member was added to the structure that would restrain a sway member. The details were passed back to the designer by the engineer and later removed from the structure.

EXERCISES

1. What are the components of the extra cost in a fast-tracked project? What are the components in the savings resulting from a reduced time in a fast-tracked project? How do you calculate the extra cost of fast-tracking a project? How do you perform a time-cost trade-off for converting a conventional project to a fast-tracked one?

2. Is it ever possible to say that fast-tracking was the essential ingredient in why a project was successful? Or does the presence of so many variables operating on a project prevent you from saying anything definite about fast-tracking? Does it remain therefore that only opinions on the contribution of fast-tracking can be held, and it is not possible to be objective on the contribution of fast-tracking to a project's success?

3. When people speak of fast-tracking, they commonly refer to overlapping the design and construction phases. Why isn't more attention focused on overlapping other parts of a project?

4. Most people say that risk increases on fast-tracked projects compared to conventional projects. What is the ingredient of risk that increases – the level of exposure (extra cost and time, unacceptable quality), or the uncertainty, or both?

PART E

A FRESH LOOK AT PROJECT MANAGEMENT

CHAPTER 16

Systems Thinking

16.1 INTRODUCTION

The conceptual framework offered by control systems theory is regarded as the most suitable for developing project management thinking.

This is developed in two ways:
- The modelling of projects as control systems, both single level and multilevel.
- The interpretation of project management as an inverse systems problem. With project changes and with project evolution, project management becomes an evolving set of inverse problems, identical in form, with non-unique solutions. With the evolution of a project, the inverse problem becomes more and more detailed. The same type of problem is being solved repeatedly throughout the project. Uniqueness is established within an optimal control systems formulation.

Projects in the following have been stripped back or simplified to their mechanical backbone. Issues relating to managing people, managing uncertainty, managing contracts and the like, which would ordinarily be superimposed on this backbone, are not considered.

16.2 SYSTEMS-SUBSYSTEMS

If a project is regarded as a system, then just as a system can be broken down into sub-systems, so a project can be broken down into subprojects. The decomposition may be continued to sub-subproject (sub-subsystem) level and beyond. (The process can also go in the other direction, from projects to program.) Decomposition can be carried out in a number of ways, for example Figure 16.1. The choice of decomposition will suit the intended project usage.

With the example decomposition of Figure 16.1a, the subsystems are all characterised similarly. With hierarchical decomposition of Figure 16.1b, successive decomposition leads to finer detail, and a change in the information type; the terms 'higher' and 'lower' are used when referring to levels in a such a decomposition. (Figure 16.1a represents a two-level hierarchy even though it is called a staged (phased) decomposition.)

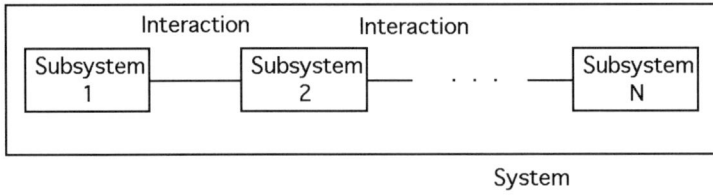

(a) Staged (phased) decomposition

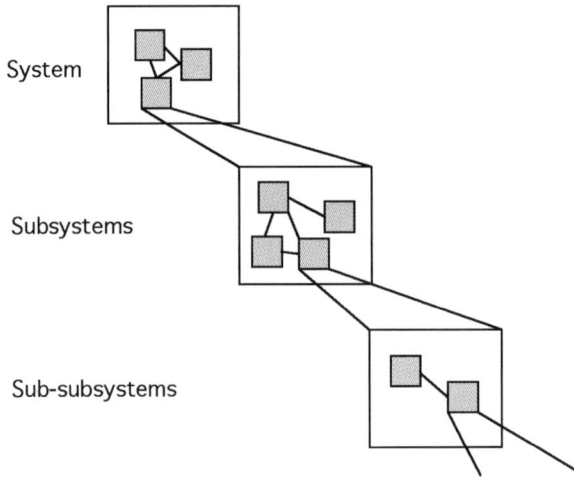

(b) Hierarchical decomposition

Figure 16.1 Example system decompositions.

Commonly, decomposition may be according to:
• Time period/project stages or phases.
• Geographic region.
• Type/nature of work.

or generally according to:
• Project.
• Subproject.
• Work package.
• Activity/task.

Subsystems combine through their interaction to give the system at the next higher level.

Subsystems, sub-subsystems, ... share the same properties and descriptors as systems, and hence are 'systems' in their own right.

16.3 SINGLE LEVEL SYSTEM

To understand systems ideas, it is convenient to start with a single level system.

Environment

The environment is defined as everything except the system. The environment affects the system by changes, and is affected by system changes. The environment might be reflected in terms of project constraints, and the environment-system interaction might be reflected in terms of initial and final (terminal) conditions.

Input and output

A system may be regarded as an input-output pair or input-output transformation (Fig. 16.2). The terms 'input' and 'output' are vector type terms, and refer to either single entities or plural entities, though sometimes the plural terms 'inputs' and 'outputs' may be used when the plural is specifically intended.

Of interest is the project (as a form of system) input and output, subproject input and output etc. Here, input is that selected by the project manager in order to get a desired project output (response or performance), which describes the project behaviour (which is typically in terms of cumulative resource usage and cumulative money usage).

Input, decision, control

Input is equivalent to a project management decision. The term 'control' is favoured here in place of both input, decision and design variable. The project manager exerts control on the project in order to bring about a desired project behaviour. The most obvious controls are resource (people, materials, equipment, ...) selections. Resources may be freely selected by the project manager, subject to any constraints being present. Where an objective/optimality criterion exists (or objectives/criteria exist), they are selected to extremise this objective.

Output, response, performance, behaviour, state

The output describes the external or observable project response, performance or behaviour. Internal system behaviour is usually given in terms of system state. By the nature

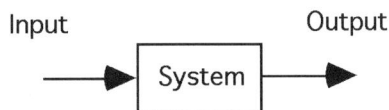

Figure 16.2 System as an input-output transformation.

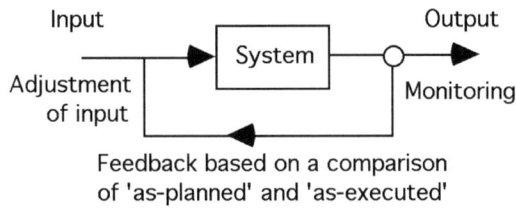

Figure 16.3 Feedback or closed loop control.

of projects, the internal and external behaviour are the same. (The observation equation of control systems theory (Carmichael, 1981) becomes 'output = state'.) There is a one-to-one transformation between output and state. The terms output and state become interchangeable when referring to projects, and there is no noise in the observation device. There are no observability (or controllability) (in the sense of Kalman) issues for projects thought of in this stripped-back simplified way. The most obvious states are the resources used so far (cumulative resources), and the cost so far (the cumulative cost). Output and state are controlled variables.

Closed loop, feedback

So-called project control involves monitoring/sensing the output ('as-executed') through time, comparing this with some 'as-planned' or pre-established baseline, and modifying the input on the basis of the difference between 'as-planned' and 'as-executed' values. This goes by the name of closed loop or feedback control (Fig. 16.3), and ensures that the project's performance goes close to that desired. It is to be compared with open loop control where the project performance may or may not go close to that desired.

    Examples of poor project control have practices that do not close the loop, close the loop without timeliness, and similar faults.

Multilevel systems

The notions of control, state, output etc all extend to subsystem, sub-subsystem etc levels, though the actual entities and their units of measurement change.

16.4 FUNDAMENTAL SYSTEMS PROBLEMS

Consider a system represented as in Figure 16.4a.
    Let the control/input be A, the model of the system B and the output C as in Figure 16.4b. The fundamental systems problems become:

Analysis:	given A and B, evaluate C
Synthesis:	given B and C, evaluate A
Investigation:	given A and C, evaluate B

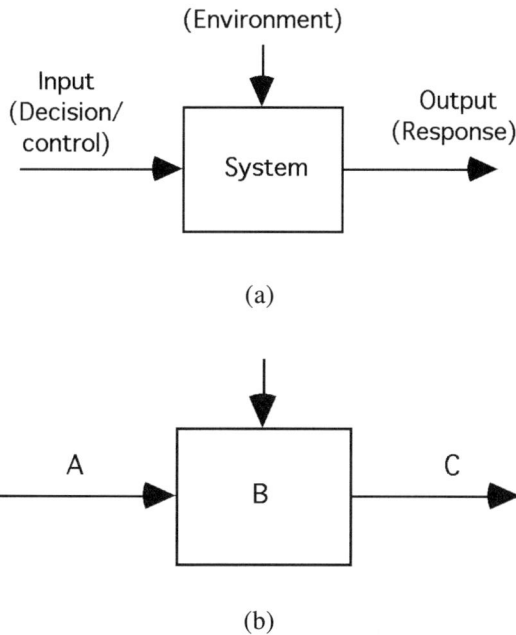

(a)

(b)

Figure 16.4 System representation.

In each, something different is known and it remains to evaluate the unknown.

Synthesis and investigation are known as *inverse problems*. In general their solution is *non-unique*, whereas the solution of the analysis problem is generally unique. Non-uniqueness in the synthesis and investigation problems might be removed through additionally including some objective/optimality criterion or equivalent.

Management is seen to fit within the synthesis categorisation. Other practices that fit within synthesis include:

- Design, optimal design.
- Optimisation.
- Optimal control theory.
- Decision theory.
- Planning.

Synthesis requires the a priori specification of a system model, that is the relationship between input and output. For (stripped-back) projects, the model is almost trivial when compared with models used in engineering and science. The input is resources (or expressed in aggregate form as money). The output is cumulative resources (or cumulative money).

The synthesis problem is essentially a converse to the analysis problem. Given a certain (desired) behaviour and model, what are the controls/decisions that produce this behaviour. Generally a certain behaviour is realisable with many controls/decisions. That is, the solution is ambiguous or *nonunique*. Further requirements such as extremisation (maximisation or minimisation) of an *objective*/optimality criterion, for example minimum time or minimum cost, are needed to make the solution unique.

*[Note: The term objective is perhaps the most popular to denote that entity by which the best solution is chosen. However different disciplines may use different terms such as (**optimality**) **criterion**, **performance index**, **payoff function**, **figure of merit**, **merit function**, **goal**, **cost function**, **design index**, **target function**, **performance measure** and **aim**.*

In this book a rigorous definition and usage of the term 'objective' is attempted. Note that the term 'objective' is used very loosely by most people. For example a project's objectives are commonly said to be the end-product of the project, or alternatively a project's objectives and its scope are talked of interchangeably. When reading the term 'objective' in non-systems work, be aware that it has a very imprecise and loose meaning, and possibly different to that given here.]

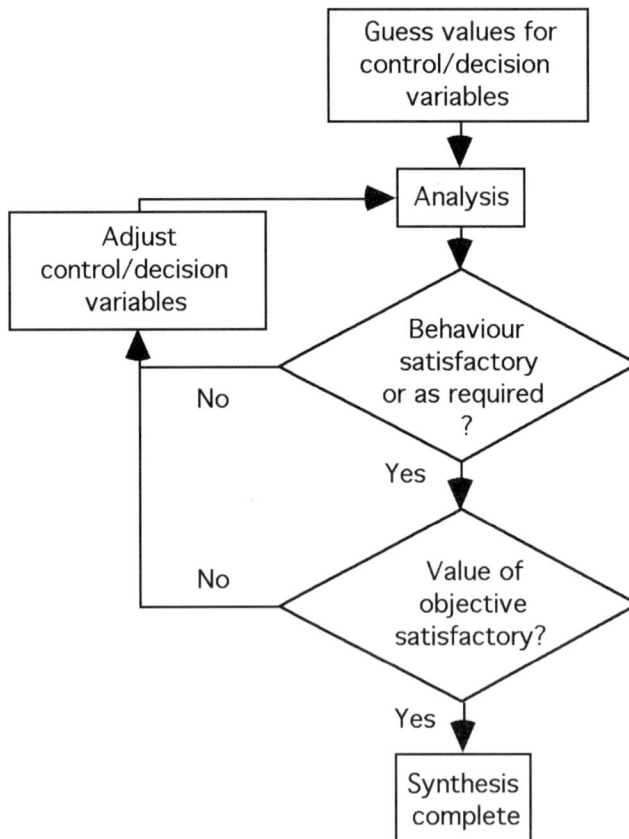

Figure 16.5 Iterative-analysis version of synthesis.

The solution of a synthesis problem could be expected to be more complicated and more difficult than the solution to the analysis problem. In some cases the analysis problem can even be solved intuitively, whereas the solution of the synthesis problem involves more rigorous methodology.

Conversion to an iterative analysis problem

Because of the degree of difficulty of solving synthesis problems compared to analysis problems, many synthesis problems are solved in an iterative-analysis fashion much like Figure 16.5.

The objective of time, cost, ... is evaluated for each set of values of the control/decision variables. Similarly the behaviour is evaluated for each set of these values. It is hoped that by adjusting the initial guesses for the control/decision variables that the behaviour resulting from the analysis will become more favourable and also the value of the objective will improve (decrease or increase as appropriate). However there is no guarantee of this, although the experience and knowledge of the person attacking the synthesis problem will usually head the iterations in the preferred direction. Where this person has synthesised such systems before, the person's first guess of values for control/decision variables may be close to 'optimum' and no iterations may be required; nevertheless the synthesis is still being approached via analysis.

16.5 MANAGEMENT

Management is a synthesis problem. As such, there are multiple solutions (decisions, or choices of control) possible. In most cases, managers are only after a satisfactory solution, or a solution that they can live with, and do not spend the additional time searching for the optimal solution. A manager may also be under time pressures to come up with quick solutions.

However managers expediently reverse the logic and deflect attention from their inability to come up with best solutions, on time pressures and pseudo 'practicality' arguments, when in fact managers do not understand the synthetic nature of their job. Managers are unaware of and do not understand the components of the synthesis problem, and so they never know where they are relative to the optimum. They are unable to vocalise or formulate the synthesis problem components; instead meaningless management jargon (Carmichael, 2002) is used as a smokescreen to hide their lack of competency. Such discussion goes to the very heart of current management knowledge being in its infancy, and current management education (read training) being superficial and low level.

Management in a systems sense

Established management procedures are iterative in nature. The iterations arise from the analysis-based mode of attack on the management problem and are not inherent in management. By suitably defining the management problem, much of the iterative process of established management procedures may be eliminated if emphasis is placed on a synthetic approach. Generally the synthesis can be done in some optimal sense, with the optimisation being performed in terms of an objective derived from imposed (often subjective) value statements. The optimum problem is then within the realm of the well delineated body of theory and techniques of optimal (control) systems or optimisation theory. In this sense the management problem is a single-level or multilevel, single objective problem. Extensions to multiple objective problems are possible.

The decision making/problem solving process

The logic of the evolution of a decision may be conveniently interpreted in the six steps of:
- *Problem definition.*
- *Value system definition.*
- *System generation.*
- *System evaluation.*
- *Selection.*
- *Action.*

The order of attack is as critical as the system modelling process and strongly influences the final decision. By providing such a framework from which to work, each step may be given a correct perspective. Feedback may occur within steps in an effort to refine the problem at any of the six steps, while a certain merging or overlapping may be noticeable between successive steps. A systematic approach to the hierarchy of steps in the decision making process generates clear thinking at each step and lends objectivity to a procedure which would otherwise be considered qualitative or intuitive.

Expressed in systems terms, the conventional notion of management decision making may be viewed as a closed loop operation of iterative modification and feedback to the analysis step (Fig. 16.6). The terminal points of the loop cycle are based, respectively:
(a) *Initial*, upon a postulated system extrapolated from experience or based on an idea, and
(b) *Final*, upon satisfaction of prearranged performance.

In short, conventional management decision making is a process of trial and error optimisation.

For many systems the step of system generation is routinely obvious with the subsystem interrelations predefined by a body of knowledge in that system's discipline. Systems are commonly divided into lower level subsystems ('subsystem delineation') to produce a tractable model and a tractable decision subproblem, although certain inconsistencies in the modelling procedure will be noticeable. In particular, there exists an interdependence of each level, requiring knowledge at a higher level when working with a lower level subsystem. No isolated systems exist. The introduction of a series of decision subproblems creates further iteration in the decision making process (that is further to that produced by an analysis-based approach). This iteration again is not inherent in decision making.

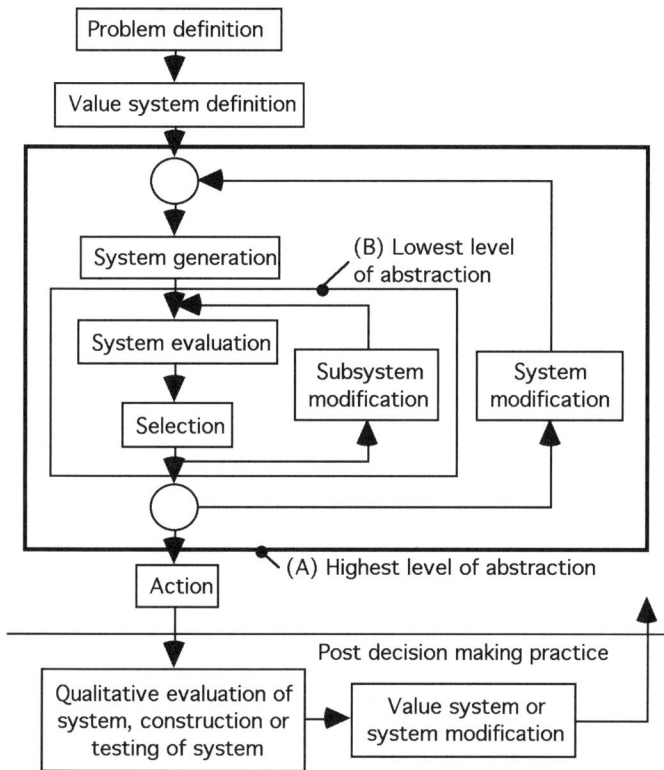

Figure 16.6 Decision making/problem solving process.

Synthetic transformations within the decision making process

The iterative nature may be partly removed from the decision making process, if for specified requirements, the system is synthesised directly to meet requirements, the operative word being 'directly'. The essential difference between the analysis-based and synthesis-based procedures is at the level of abstraction adopted in the computations. (The terminology 'level of abstraction' is used in the sense relating to the quantity of a priori data assumed.) Analysis-based techniques impose a total system configuration *ab initio*, while the system emerges from any given level of abstraction as a natural consequence of the direct synthetic treatment. Presumably the extreme generality that may be attained in the direct case would involve little or no a priori knowledge of the emerging system – refer level of abstraction (A) in Figure 16.6. However, for a solution of practical significance, certain leading properties of the system configuration are best assumed – the corresponding level of abstraction is intermediate between levels (A) and (B) in Figure 16.6. The choice of abstraction level on which the decision maker chooses to work would be a balance between his/her expertise-based judgements and desired computation load. A synthesis-type treatment can only proceed where certain of the system properties remain free and adjustable. It is apparent that a synthesis-type format to decision making is the fundamental and at the same time more rational approach.

The theory of optimal control systems

Generally one desires to synthesise a system which is optimal in a certain sense; an objective, criterion, index, ..., resulting from an imposed value statement, is implied. Optimum systems are of central concern in optimal control theory which exploits the synthetic nature of the problem. It is the philosophy of this theory that is found most useful in the decision problem. The philosophy rests on very broad grounds, typical of techniques in systems theory, only conversing in the entities state and control (and sometimes response), which take on very definite meanings in the decision problem.

In simple terms, synthesis is thus equivalent to choosing the controls throughout the system; optimal synthesis or optimal control selects the controls so as to extremise some objective. In addition, supplementary constraints are also usually present. It is the theory of control that is concerned with the mathematical formulation of laws for the control of systems.

Optimal control theory in some engineering branches has elevated the 'art' of decision making to a status approaching a systematic and exact 'science'. This has occurred despite the ever present yet necessary 'expertise-based' judgements, which recognise the existence in all decisions of certain intangible quantities that defy precise mathematical statements. This theory offers the same advantages to management.

16.6 THE OPTIMAL CONTROL PROBLEM

The problem of developing an optimal system is one of finding an admissible control such that the system functions with the objective extremised. The objective is expressed analytically as a function with, in general, both state and control variable arguments. Certain physical, economic, operational, engineering and other constraints may be present, restricting the control choice. This choice may also be simplified if the search is confined to certain classes of systems. The problem is thus one of choosing the system control such that the system operates in some best way, while observing the constraints present.

Problem components

The formulation of an optimal control problem requires the following components:
(i) *A model of the system to be controlled*
     This is often a constitutive equation, together where applicable with initial and final (terminal) conditions, ideally expressed in a standard form. It characterises the system and enables the effect of alternative controls on the system to be predicted. Chapter 17 gives example system equations (model), including a standard state equation form.

(ii) *The constraints upon the decision making*
Constraints limit the range of permissible solutions and fix many of the system properties.

(iii) *The demands presented to the system in the form of an objective (criterion, index, goal, ...)*
The objective is derived from a value statement. To evaluate possible alternative
solutions, a scalar index is introduced. The problem is to determine the control that
gives the least or greatest value of this index.

Solution controls are said to be *feasible* or *admissible* if they satisfy the system model
and are within the permissible bounds as defined by the constraints. Where a range of
admissible solutions exists the problem is considered well posed. The objective provides
the criterion by which the optimal control is chosen from the set of admissible controls in
order that the constraints are satisfied in a best manner.

Using a systems – state and control – foundation, superficially different synthesis problems can be shown to share a common mathematical basis, leading to common solution
techniques.

The optimal control problem of Chapter 17 is for the single level, single objective case.
Multilevel and multiobjective cases are possible.

Constraints

Constraints influence the solution characteristics by isolating admissible solutions, from
all possible solutions, and give meaning to the choice of the optimal system. Constraints
may be defined throughout time, or at points or on intervals of time, and are given in the
form of inequalities or equalities.

In a sense, system models in the form of equations may be regarded as equality constraints. The system is constrained to belong to the class of systems whose equations are
of this form. Initial and final (terminal) conditions may be likewise treated as equality
constraints.

Constraints typically restrict the values taken by the state and control variables. The
range of possible values that the states and controls may assume is reduced to a set of
admissible values.

Objective

Objectives provide the means of quantitatively assessing alternative solutions. The
solutions are only optimal in the sense of the objectives which follow from the problem
statement, although computational tractability reasons may warrant introducing alternative, simpler objectives. The latter objectives obviously lead to suboptimal decisions with
respect to the original objectives.

Optimality implies an extremisation requirement on some measure J, the objective.
In general this measure is a function of both state and control variables as well as time
coordinates, and is a scalar quantity.

The objective J may be thought of as assigning a unique number to each admissible
solution. The optimum J is chosen from the many admissible values of J. Alternatively

J may be considered as a function in which the controls play the role of the independent variables. The objective derives from an imposed value system, the correct identification of which remains essential for a meaningful problem. For a mathematical formulation, quantitative measures must replace qualitative measures (in subjective value systems).

Suboptimal control assumes an additional role to that mentioned above. In particular the implementation of the optimal control may be inadmissible for engineering, economic or other reasons (that is other constraints not allowed for in the mathematics of the problem). Knowing the optimal control enables the implementation of a suboptimal form with a full understanding of the consequences of such action. In this sense the optimal control serves as a benchmark by which alternative controls may be evaluated.

Without loss of generality, minimisation is commonly implied in all optimisation studies. It will be appreciated that any problem in maximisation may be conveniently treated as a problem in minimisation by means of a suitable negative transformation:

$$\max(-J) = -\min(J).$$

Probabilistic systems

For probabilistic arguments, the general objective J is now a probabilistic quantity, and hence an unsuitable measure. A suitable deterministic measure, over which the minimisation may be carried out, is the expected value or first moment (in a probabilistic sense) of J, $E\{J\}$, where $E\{\bullet\}$ denotes the expectation operation. The expectation operation may be visualised as taking the average of the objective evaluated for each of the possible values of its arguments.

In general, this expected value of the probabilistic measure J, is used as the objective. However, in certain applications a measure or index of reliability may be relevant; that is extremising the index may relate to minimising the probability of the system exceeding (both positive and negative senses implied together or singly) a particular limit state, or maximising the probability of non-exceedance in order that the system attains a maximum level of reliability. Notice that this is a different situation to the one in which a system is synthesised for a given reliability (the probability of the state exceeding a given limit state is prescribed). Reliability in this context is a constraint.

Multiple objectives

Several objectives (resulting from multiple requirements) expressed for the one problem in general lead to different solutions (values of control variables). In general the solutions do not coincide and hence the existence of more than one objective simultaneously is worrying for a meaningful problem. Auxiliary conditions, equivalent to constraints, may however, coexist with the objective. Adaptations of Lagrangian multiplier and weighting function concepts may also be employed for a solution. Alternatively trade-offs or adjustments may be made between the several objectives.

Alternative terminology

Some writers refer to the totality of these three problem components as a model of the optimisation problem or a model of the decision making process; this terminology is not followed here.

The term *restraint* is sometimes used by writers instead of constraint; this terminology is not followed here.

The term objective is perhaps the most popular to denote that entity by which the best solution is chosen. However different disciplines may use different terms such as (*optimality*) *criterion, performance index, payoff function, figure of merit, merit function, goal, cost function, design index, target function, performance measure* and *aim*.

Example

Consider an operation involving equipment.

The designer of this operation selects the number and type of equipment. These may be referred to as control variables.

In selecting these controls the designer is trying to achieve for the operation:

• Appropriate utilisation (proportion of time working) of the equipment.
• Desired output or productivity (ratio of output to input).

These may be referred to as state variables.

As well, the designer may be trying to achieve least cost per production.

For such operations a suitable system model might be a queueing/waiting line theory model or a simulation model. (Carmichael, 1987) Subsystem models may relate to equipment characteristics.

Constraints may relate to production levels, durations and so on.

The designer guesses or uses experience or industry knowledge to establish numbers and types of equipment. The operation is analysed, and utilisations, productivities and costs are obtained. Numbers and types of equipment are adjusted until the desired operation characteristics are obtained.

The designer is attempting to come up with an equipment configuration that is least cost per production while satisfying any constraints present.

The problem may be simplified by fixing the type of equipment or the designer may be constrained to using available equipment.

In terms of an optimal control problem, the components are:

Model	– Queueing or simulation model; equipment performance characteristics.
Constraints	– Related to time and production.
Objective	– (minimum) Cost per production.

The controls are selected to minimise the objective while satisfying the model and constraints.

16.7 OPTIMISATION TECHNIQUES

Optimisation attempts to find the extremum (maximum or minimum) of the objective while satisfying any constraints and the system model. It may be approached through consideration of the mathematics alone, by ignoring sophisticated mathematics and using numerical evaluation, by nonrigorous mathematics, by 'gut feel' considerations, or combinations of these.

'Gut feel' considerations

If the synthesis problem is sensibly posed and relates to a discipline in which the decision maker has expertise, frequently something close to the optimum can be obtained by 'gut feel' considerations.

Numerical approaches

Numerical approaches simply evaluate the objective for a range of values of its arguments. That set of values which gives the lowest (or highest) value of the objective is taken as the optimum.

Calculus

There exists the calculus of extrema for finding minima, maxima and points of inflection of analytical functions (objectives). These points are obtained by setting the first derivative of the function to zero (necessary conditions for extrema). Checking the sign of the second derivative of the function establishes the type of extrema found.

Extensions using Lagrange multipliers can deal with the presence of constraints.

Pontryagin's maximum principle

A generalisation of the calculus and Lagrange multipliers to the dynamic system model case leads to Pontryagin's maximum principle (or the calculus of variations). The resulting necessary conditions of optimality are a set of simultaneous equations in the state and costate (Lagrange multipliers) variables.

Dynamic programming

Similar dynamic problems may be solved using dynamic programming based on Bellman's principle of optimality.

The technique also extends to sequential or staged (phased) systems, and to decision problems generally.

Mathematical programming

Where the decision problem components are a collection of algebraic relationships, mathematical programming can be applied.

Linear programming gives the solution where the relationships are linear.

Non linear programming techniques apply where the relationships are nonlinear. Innumerable methods have been proposed and include:
• Geometric programming.
• Quadratic programming.
• Search methods.
• Gradient methods.
• Function approximation.
• Penalty function methods.
• Method of feasible directions.

To demonstrate their optimality, sometimes reference is made to the Kuhn-Tucker conditions which are necessary for an optimum for the constrained optimisation case.

16.8 PLANNING

Planning envisages *how the job will be done, in what order* and *with what resources* (people, materials, equipment). When it is to be done by, when the resources are needed etc is usually referred to as scheduling; planning and scheduling are very closely linked, so much so that many people include timing matters within a larger definition of planning.

Planning involves *creativity* as well as *experience*, and can be a demanding mental activity.

Planning is commonly looked at in relation to:
• The associated timing of the method and order of the work (referred to as *scheduling*).
• The use of money (*budgeting* or *financial planning*) as a common unit of measurement of resource usage.
• The use of resources – people (*human resources planning*), materials and equipment (*resource planning*).

There are a number of useful *techniques* (such as the critical path method, CPM) to assist the planner analyse, organise and present the information. Presentations, in the form of charts, graphs and so on, enable the *communication* of the plan.

Example

Planning establishes how work will be done, in what order and with what resources.

Consider the simpler (subsystem) problem where it has already been largely established how the work will be done and with what type of resources. As such, the controls available to the planner are the order or sequence of the work including when particular work items will be done.

This establishes when resources are used, or if expressed in monetary units, when money is used.

What the planner is trying to influence is the utilisation of the resources (state variables), and thereby the duration of the work.

There may be constraints on costs, duration of the work and resource availability.

Ultimately what the planner might be trying to achieve is a work sequence of minimum cost or minimum duration or a compromise between these two objectives.

A suitable system model is the time-scaled critical path network, linked bar chart, multi activity chart or flow process diagram, with connections to resource histograms and cash flow diagrams.

As the planner adjusts the sequence of work, so resource utilisations, alternative cash flows, and work durations evolve. By a series of iterations, the planner settles on one particular sequence of work.

Using an optimal control approach, the components to the problem are:

Model   – Time-scaled network, linked bar chart, ...
Constraints – Related to work duration, costs, resource availability.
Objective   – (minimum) Cost and/or (minimum) time.

16.9 PROJECT MANAGEMENT

Project management translates to a sequence of decisions within an hierarchical framework.

Upper hierarchy level

The upper level of the hierarchy occurs in starting a project off (Chapter 5). Here there are two main synthesis problems (Chapter 5, Fig. 5.1):
- 'End-product' – objectives and constraints are established for the end-product, and the solution of the synthesis problem leads to the selection of the end-product.
- 'Means to End-product' – objectives and constraints derive from a consideration of the end-product objectives and constraints, as well as the project itself, and the solution of the synthesis problem leads to the selection of the project scope.

Lower hierarchy level

Multiple lower level synthesis problems over time are present while progressing a project (Chapter 9, Section 9.3). Here the higher level value systems that lead to the establishment of the higher level objectives and constraints transfer directly. The solutions to the synthesis problems at this lower level lead to decisions on order of work, resourcing etc.

CHAPTER 17

Staged (Phased) Decomposition of Projects

17.1 INTRODUCTION

A project may be decomposed into stages or phases. If the stages are sequential (that is, the staging is based on a time subdivision; each stage represents a period of time) then the decomposition will look like Figure 17.1. Each stage is a subproject. There may be hold or approval points between each stage.

In Figure 17.1,
N   number of stages
$x(k)$ state, $k = 0, 2, ..., N$
$x(0)$ initial state
$x(N)$ final state
$u(k)$ control at stage k, $k = 0, 1, 2, ..., N-1$

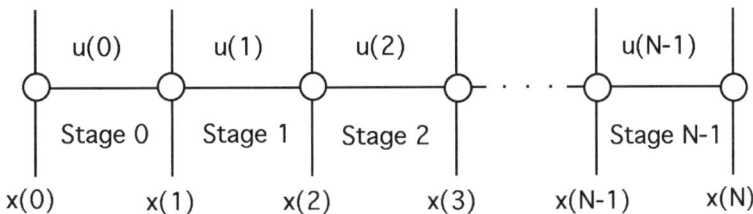

(a) 'Activity-on-arrow' form.

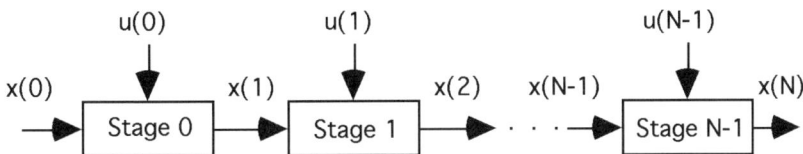

(b) 'Activity-on-node' form.

Figure 17.1 Staged (phased) decomposition of a project.

State and control are vector functions, $x = (x_1, x_2, ..., x_n)^T$; $u = (u_1, u_2, ..., u_r)^T$

Associated with each stage is an objective function $J(k)$, $k = 0, 1, 2, ..., N-1$ (See the earlier definition of the term objective used in this book. It is not used in a loose lay person, dictionary or management text sense.)

Note that the stage output and the state are the same in this formulation.

Such a formulation is amenable to the application of multistage decision problem techniques such as dynamic programming. There, an optimisation problem is broken down into subproblems associated with each stage. The optimisation problem is solved by successively going through each stage, from start to finish (or finish to start). Associated with each stage is an objective function which evaluates the worth of each stage decision (control) made from a range of possible alternative decisions (controls). The state represents the link between successive stages; it carries forward information on the system behaviour, such that the state $x(k)$ incorporates information from stages 0 to k-1.

What constitutes the state and control is perhaps the hardest thing to establish when setting up such a multistage decision formulation. The state differs from application to application.

The state at any stage is the minimal amount of information needed to completely determine the behaviour (state) of the system for all other following stages, for any given control. To extend the concept of state to stochastic systems, the state at any stage is regarded as the information that uniquely determines the probability distributions of behaviour (state) at all other following stages. That is, a discrete equivalent of a Markov process. (Carmichael, 1981)

The state is information needed from all preceding stages in order to make a decision in the current stage without reference to previous decisions made. It represents status information on the project at any point in time.

Some reasonable choices of state and control in the context of projects follows.

Individual resource viewpoint

state(k):　cumulative resource usage; the resource usage of stages up to and including stage k-1

control(k):　resource level chosen at stage k

Practitioners have unknowingly been using state ideas when employing cumulative resource usage plots.

The number of states and controls reflects the number of resources. Resources here refer to people, equipment, materials and similar. Many resource allocation problems have similar state and control definitions.

This makes for a very high dimensional problem. The dimension can be reduced if resources are expressed in a common unit of money and aggregated.

Aggregated resource, cost viewpoint

state(k):　cumulative cost; the expenditure of stages up to and including stage k-1

control(k):　expenditure level chosen at stage k

Practitioners have unknowingly been using state ideas with their usage of S (cumulative cost) curves.

An alternative control

It may be preferable to think of the control as a change in resource levels or a change in expenditure between stages, rather than absolute resource levels used at each stage. This is not adopted here.

State equations

Let x(k) be the cumulative resource usage/cost. If cumulative resource usage is chosen (rather than cumulative cost), then x is a vector of resource usages, as large as there are resources being considered. The system equations (model) in state equation (state transformation) form become

x(k+1) = x(k) + u(k)

where
x(k)   cumulative resource usage/cost for stages 0 to k-1
u(k)   resource usage/expenditure in stage k

The initial state x(0) = 0, but the final state x(N) is not fixed.
   The above state equations are characteristic of discrete (data) systems.
   State trajectories may look something like Figure 17.2.

In order to track project time, a relationship between resource usage/expenditure and time taken is needed. This implies some productivity type relationship, for example Figure 17.3. Or utility curves (activity cost-time data) (Carmichael, 1989), which are usually drawn with the axes interchanged to that of Figure 17.3, could be used.

From Figure 17.3,

$$T(k) = a - \frac{a}{b} u\,(k)$$

where
T(k)   duration of stage k
a, b   constants
u(k)   resources used/expenditure in stage k

Objective/optimality criterion

There are a number of possible choices available to the project manager. The most obvious would be minimum resource usage/expenditure and minimum time.

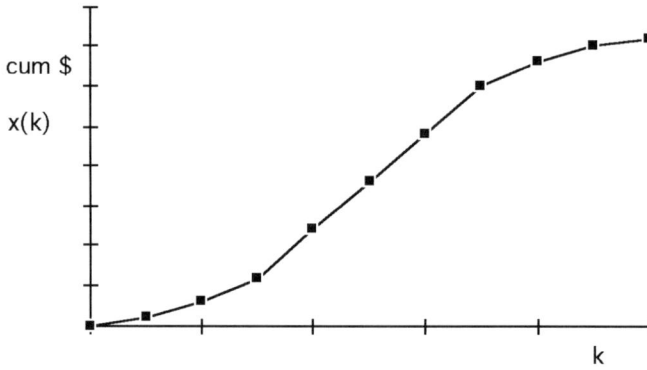

(a) Money usage.

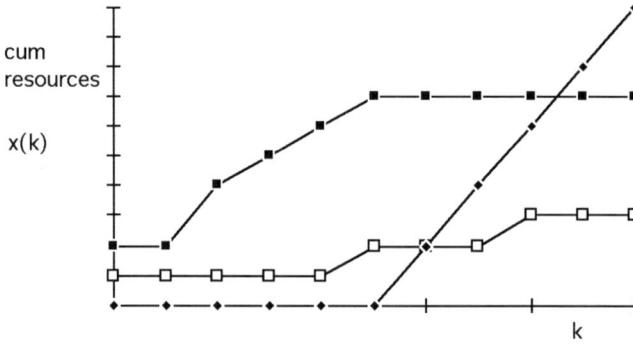

(b) Resource usage.

Figure 17.2 Example state trajectories.

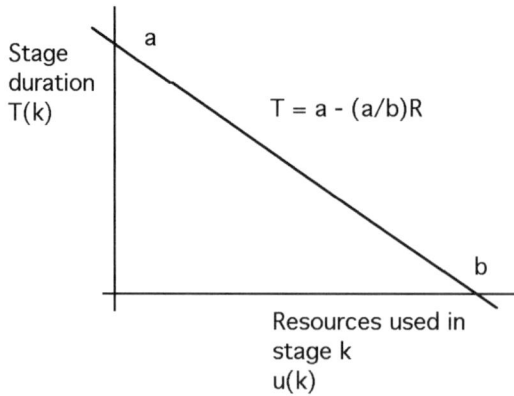

Figure 17.3 Example relationship between resource usage/expenditure and time taken.

Minimum resource usage/expenditure. For each stage, $J(k) = u(k)$, and

$$\min J = \min \sum_{k=0}^{N-1} u(k)$$

Minimum time. For each stage, $J(k) = T(k)$, and

$$\min J = \min \sum_{k=0}^{N-1} T(k)$$

This reduces to

$$\min \sum_{k=0}^{N-1} a - \frac{a}{b} u(k) = \min \sum_{k=0}^{N-1} -u(k) = \max \sum_{k=0}^{N-1} u(k)$$

That is, minimum time corresponds with maximum resource usage/expenditure (assuming Figure 17.3 applies).

Constraints

Typically there may be constraints on the overall expenditure and project duration,

$x(N) \leq$ some limit

$$\sum_{k=0}^{N-1} T(k) = \sum_{k=0}^{N-1} a - \frac{a}{b} u(k) \leq \text{some limit}$$

The second constraint translates to a constraint on the controls.

There may also be restrictions on resource usage/expenditure throughout the project,

$u(k) \leq$ some limit $\qquad\qquad\qquad k = 0, 1, 2, ..., N-1$

Linearity

Because of the linear state equations, the optimal solution may turn out to be *singular*, possibly even *bang-bang* in nature (Carmichael, 1981). This will depend on the nature of the objective/optimality criterion and constraints chosen.

17.2 A GENERALISATION

This development may be shown to fit within a general optimal control problem statement. (Carmichael, 1981)

System model

The general form of the state equations given in the previous section is,

$x(k+1) = F[x(k), u(k), k]$ $k = 0, 1, ..., N-1$

where

$x(k) = [x_1, ..., x_n]^T$ is an n-valued vector of the state at stage k

$u(k) = [u_1, ..., u_r]^T$ is an r-valued vector of controls at stage k

$F = [F_1, ..., F_n]^T$ is a nonlinear vector function

   For given u and initial and final (terminal) conditions on x, x is completely defined by the state equations.

   They represent a sequence of transitions from the k'th to the (k+1)'th state, k = 0, 1, ..., N-1. The state is assumed to have Markov properties; that is, the state at k+1 depends only on the immediately previous state x(k) and control u(k).

Remaining optimal control problem components

Objective/optimality criterion

A general objective, incorporating the earlier examples, can be written as,

$$\min J = g[x(k)]\big|_{k=0}^{k=N} + \sum_{k=0}^{N-1} G[x(k), u(k), k]$$

where

G, g are scalar single-valued functions of their respective arguments.

Constraints

Constraints influence the solution by isolating admissible solutions from all possible solutions. Constraints may be given in the form of equalities or inequalities. Constraints may exist on both state and control variables.

Common project constraints include limits on expenditure and limits on resource usage.

The problem

The problem is to determine an admissible sequence of controls u(k), k = 0, 1, ..., N-1 satisfying the system (state) equations, initial and final (terminal) conditions on the state, and any constraints, and minimising J. û(k), k = 0, 1, ..., N-1 is termed the optimal sequence or policy.

The linear-quadratic case

The case involving linear system equations and quadratic objective/optimality criterion has a particularly neat solution. (Carmichael, 1981)

Determinism

The above is a deterministic formulation. A related stochastic formulation can be given.

17.3 DYNAMIC PROGRAMMING

Dynamic programming, as originally devised by Bellman, was a technique for solving variational and multistage decision problems, but was subsequently extended to other problems including optimal control problems. At its base is the *principle of optimality* and *Markovian* properties of the state. The principle states: '*An optimal policy has the property that whatever the initial state and initial decision, the remaining decisions must constitute an optimal policy with regard to the state resulting from the first decision.*' The words decision and control may be interchanged. The principle of optimality may be proved by contradiction. The resulting recurrence relation expressing optimality is referred to as *Bellman's equation*. For continuous systems, the technique is logically equivalent to the *maximum principle of Pontryagin* and the classical *calculus of variations*. (Carmichael, 1981)

Dynamic programming is best suited to systems with a serial structure or staging, but can be applied to systems with parallel arrangements.

Dynamic programming is an approach to an optimisation problem rather than a general solution algorithm. The approach uses an embedding procedure whereby the original problem is replaced by a sequence of smaller problems. It is noted that there are limitations to its practical application but it is potentially powerful and provides a neat treatment of certain optimisation problems.

Formulation

Assume g is given for $k = N$ only. Assume initial conditions, $x(0)$ only.

The appropriate relationships are developed in the reverse time direction.

Dynamic programming reduces this minimisation problem of N vectors $u(0)$, ..., $u(N-1)$ to a sequence of N minimisations over a single vector. This is possible as the given optimisation problem is embedded into a wider class of problems. Define,

$$J_{N-1} = G[x(N-1),u(N-1),N-1] + g[x(N)]$$
$$J_{N-2} = G[x(N-2),u(N-2),N-2] + J_{N-1}$$
etc

and the optimal return functions,

$$S_{N-1} = \min_{u(N-1)} J_{N-1}$$

$$S_{N-2} = \min_{u(N-2),u(N-1)} J_{N-2}$$

...

$$S_{N-j} = \min_{u(N-j),...,u(N-1)} J_{N-j} \qquad\qquad j = 3,4,...,N$$

Here S_{N-j} is the minimum value of the criterion associated with the optimal j-stage discrete process with initial state $x(N-j)$. Applying the principle of optimality, gives the recurrence formula

$$S_{N-j}[x(N-j)] = \min_{u(N-j)}\{G[x(N-j),u(N-j),N-j] + S_{N-j+1}[x(N-j+1)]\}$$

where

$$x(N-j+1) = F[x(N-j),u(N-j),N-j]$$

The sequential solution procedure starts with $j = 1$, finding $S_{N-1}[x(N-1)]$ and $\hat{u}(N-1)$, and so on for $j = 2, 3, \ldots, N$ giving a sequence of pairs $S_{N-j}[x(N-j)]$ and $\hat{u}(N-j)$ ending with $S_0[x(0)]$ and $\hat{u}(0)$. At each stage S_{N-j} replaces S_{N-j+1} in 'memory'. The sequence $\hat{u}(0)$, $\hat{u}(1)$, ..., $\hat{u}(N-1)$ constitutes the optimal control sequence or policy and $S_0[x(0)]$ is the value of the objective/criterion. The states $x(1), \ldots, x(N)$ can be found in succession from the system equations since the controls are now known.

This solution procedure can in certain circumstances lead to a closed form solution or it can be used as a computational algorithm. Note that for large n, the amount of computer storage required is large and some approximating process is needed. For a numerical solution, the optimum at each stage can be conveniently done by a search procedure in conjunction with interpolation schemes.

All minimisations are assumed to be carried out subject to any constraints on the problem. Some constraints may be handled directly and numerically simplify the computations. Alternatively the constraints can be adjoined to the objective/criterion by means of Lagrange multipliers to give an augmented objective/criterion. The problem then is to find the control and the multipliers which minimise the augmented criterion.

The problem above is decomposed additively. The objective function is composed of the sum of components from each stage. Other decompositions are possible within a dynamic programming framework.

17.4 PONTRYAGIN'S PRINCIPLE

Pontryagin's maximum principle gives a set of necessary conditions for an optimum. (Carmichael, 1981)

Necessary conditions

Consider x given at $k = 0$ and $k = N$, and constraints on u. A Hamiltonian function is defined by adjoining the system equations to the objective/optimality criterion by means of Lagrange multipliers (costate variables, adjoint variables) $\lambda = [\lambda_1, \lambda_2, \ldots, \lambda_n]^T$,

$$H[x(k),u(k),\lambda(k+1),k] = G[x(k),u(k),k] + \lambda^T(k+1)F[x(k),u(k),k]$$

Then $\hat{\lambda}$ satisfies the following costate equations, and initial and final (terminal) conditions,

$$\hat{\lambda}(k) = \frac{\partial H[\hat{x}(k), \hat{u}(k), \hat{\lambda}(k+1), k]}{\partial \hat{x}(k)}$$

$$\left[\hat{\lambda} - \frac{\partial g}{\partial \hat{x}}\right]_{k=0}^{k=N} = 0$$

and

$$H[\hat{x}(k), \hat{u}(k), \hat{\lambda}(k+1), k] = \min_{\substack{u \text{ constraints}}} H[x(k), u(k), \lambda(k+1), k] \quad k = 0, 1, ..., N\text{-}1$$

Where u(k) is unconstrained,

$$\frac{\partial H[x(k), u(k), \lambda(k+1), k]}{\partial u(k)} = 0 \quad k = 0, 1, ..., N-1$$

As well, there are the original system (state) equations with initial and final conditions. These necessary conditions constitute a discrete two point 'boundary' value problem, and are solved by numerical computation.

The state equations and costate equations (involving λ) are referred to as canonical equations.

More general initial and final conditions can be considered.

17.5 LOWER LEVEL DECOMPOSITION OF PROJECTS

Stages may be decomposed to finer detail, right down to activity level. Consider, for example, the network (activity-on-arrow precedence diagram) of Figure 17.4 (Carmichael, 1989). As for Figure 17.1 above, arrow diagrams may be easier to work with than activity-on-node precedence diagrams. Interestingly, practitioners tend to favour linked bar charts when representing project programs, and linked bar charts are no more than time-scaled arrow diagrams.

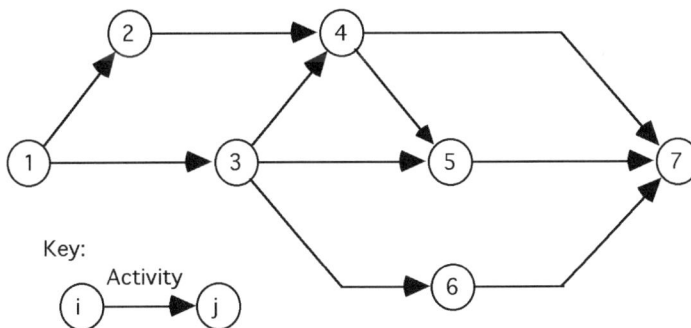

Figure 17.4 Example network (activity-on-arrow).

The project may be subdivided into subprojects or stages. The choice where the sub-division occurs depends on the user and the nature of the project, but in principle any subdivision is possible.

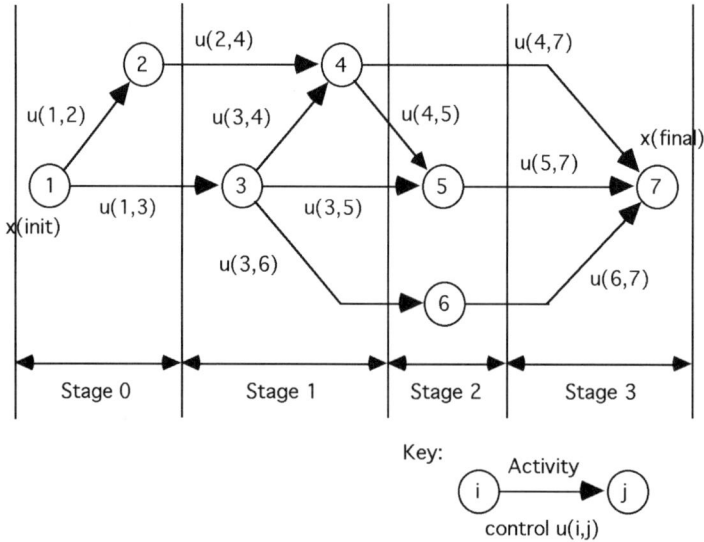

(a) 'Activity-on-arrow' form.

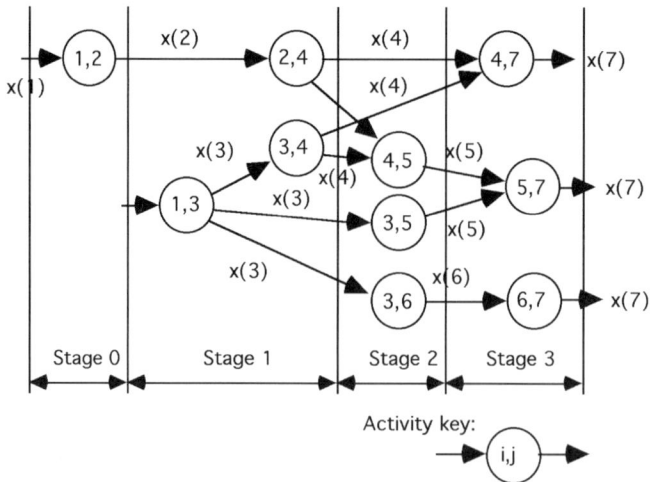

.(b) 'Activity-on-node' form.

Figure 17.5 Example subdivision of network.

Consider, for example, the subdivision shown in Figure 17.5, where vertical lines have been drawn through the network, though the approach does not require subdivisions to be vertical. The portions between subdivisions have been called stages, but could also be called subprojects or phases. Controls u(i,j) are shown adjacent to each activity (i,j).

The inputs and outputs (cumulative resource usage/expenditure) for each stage are as follows, working from the activity-on-arrow diagram:

Stage 0

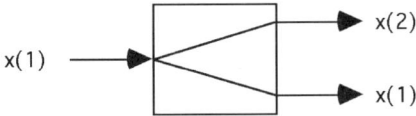

$x(2) = x(1) + u(1,2)$
$x(1) = x(1)$

Stage 1

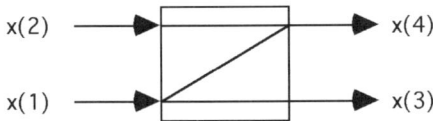

$x(4) = x(2) + u(2,4) + x(3) + u(3,4)$
$x(3) = x(1) + u(1,3)$

Stage 2

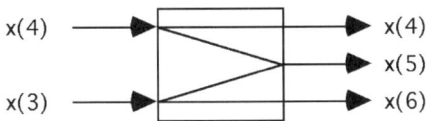

$x(4) = x(4)$
$x(5) = x(4) + u(4,5) + x(3) + u(3,5)$
On eliminating a redundancy,
$x(5) = x(4) + u(4,5) + x(3) - x(3) + u(3,5)$
$x(6) = x(3) + u(3,6)$

Stage 3

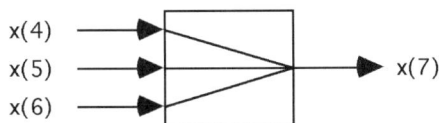

$x(7) = x(4) + u(4,7) + x(5) + u(5,7) + x(6) + u(6,7)$
On eliminating redundancies,
$x(7) = x(4) + u(4,7) + x(5) - x(4) + u(5,7) + x(6) - x(3) + u(6,7)$

Collectively the system equations may be written,

$$
\begin{bmatrix}
x(1) \\
x(2) \\
x(3) \\
x(4) \\
x(5) \\
x(6) \\
x(7)
\end{bmatrix}
=
\begin{bmatrix}
x(1) \\
x(1) \\
x(1) \\
x(2)+x(3) \\
x(4) \\
x(3) \\
-x(3)+x(5)+x(6)
\end{bmatrix}
+
\begin{bmatrix}
0 \\
u(1,2) \\
u(1,3) \\
u(2,4)+u(3,4) \\
u(4,5)+u(3,5) \\
u(3,6) \\
u(4,7)+u(5,7)+(6,7)
\end{bmatrix}
$$

The system equations are linear. The individual activity controls can be grouped in any desired combinations, but presumably the groupings in the above system equations would be leading contenders. The dimension of the problem can be reduced by selecting fewer stages and showing some discrimination in the selection of stages. Consider, for example, Figure 17.6.

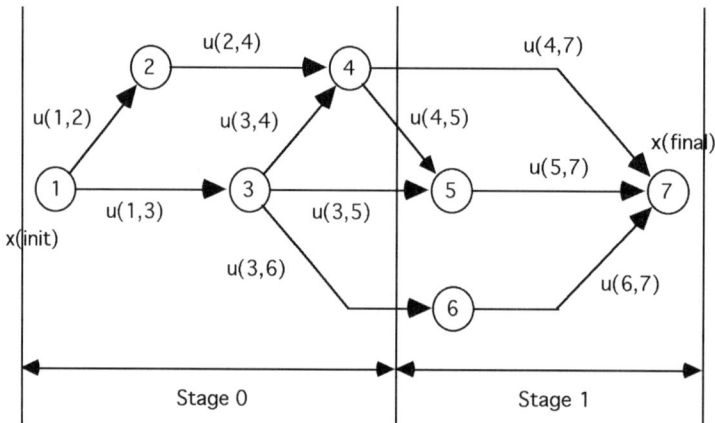

Figure 17.6 Example subdivision of network.

The inputs and outputs (cumulative resource usage/expenditure) for each stage are as follows:

Stage 0

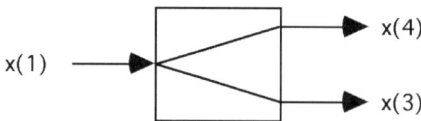

$x(4) = x(1) + u(1,2) + u(2,4) + x(3) + u(3,4)$
$x(3) = x(1) + u(1,3)$

Stage 1

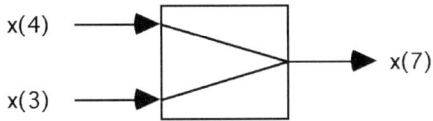

$x(7) = x(4) + u(4,7) + u(4,5) + u(5,7) + x(3) + u(3,5) + u(3,6) + u(6,7)$
On removing redundancies,
$x(7) = x(4) + u(4,7) + u(4,5) + u(5,7) + x(3) + u(3,5) - x(3) + u(3,6) + u(6,7)$

Collectively the system equations may be written,

$$
\begin{bmatrix} x(1) \\ x(3) \\ x(4) \\ x(7) \end{bmatrix} = \begin{bmatrix} x(1) \\ x(1) \\ x(1)+x(3) \\ x(4) \end{bmatrix} + \begin{bmatrix} 0 \\ u(1,3) \\ u(1,2)+u(2,4)+u(3,4) \\ u(4,7)+u(4,5)+u(5,7)+u(3,5)+u(3,6)+u(6,7) \end{bmatrix}
$$

or

$$
\begin{bmatrix} x(1) \\ x(3) \\ x(4) \\ x(7) \end{bmatrix} = \begin{bmatrix} x(1) \\ x(1) \\ x(1)+x(3) \\ x(4) \end{bmatrix} + \begin{bmatrix} 0 \\ u1 \\ u2 \\ u3 \end{bmatrix}
$$

where,

$u1 = u(1,3)$
$u2 = u(1,2) + u(2,4) + u(3,4)$
$u3 = u(4,7) + u(4,5) + u(5,7) + u(3,5) + u(3,6) + u(6,7)$

The solution to the associated optimisation problems might best be done using mathematical programming or dynamic programming (Carmichael, 1981). Linear programming may be preferred for computational reasons.

References and Bibliography

Al-Bahar, J.F. & Crandall, K.C. (1990). Systematic Risk Management Approach for Construction Projects. *ASCE Jnl of Const. Eng. and Mgt., 116*(3), 533-546.

ASCE (1993). *Journal of Management in Engineering, 9*(4), 303-304.

Barrie, D.S. & Paulson, B.C. (1992). *Professional Construction Management.* McGraw-Hill, New York, 3rd ed.

Bowen, J. (1987). *The Macquarie Easy Guide to Australian Law.* The Macquarie Library Pty Ltd, Sydney.

Carmichael, D.G. (1981). *Structural Modelling and Optimization.* Ellis Horwood (John Wiley and Sons), Chichester.

Carmichael, D.G. (1987). *Engineering Queues in Construction and Mining.* Ellis Horwood Ltd (John Wiley and Sons), Chichester.

Carmichael, D.G. (1989). *Construction Engineering Networks.* Ellis Horwood (John Wiley and Sons), Chichester.

Carmichael, D.G. (1996). Flat Organisational Structures. *Journal of Project and Construction Management, 2*(2), 61-68.

Carmichael, D.G. (1997). Reengineering and Work Study. *Journal of Project and Construction Management, 3*(2), 95-107.

Carmichael, D.G. (2000). *Contracts and International Project Management.* A.A. Balkema, Rotterdam.

Carmichael, D.G. (2002). *Disputes and International Projects.* A.A. Balkema, Rotterdam.

Currie, R.M. (1959). *Work Study.* Pitman, London.

Dell'Isola, A.J. (1982). *Value Engineering in the Construction Industry.* Construction Publishing Corp., New York.

Hall, A.D. (1962). *A Methodology for Problem Solving.* Van Nostrand, New Jersey.

Institute of Building (1979). *Project Management in Building.* The Institute of Building, Ascot.

International Labour Office (ILO) (1969). *Introduction to Work Study.* ILO Geneva, 2nd ed.

Kerzner, H. (1989). *Project Management.* Van Nostrand Reinhold, 3rd ed., New York.

New South Wales Premier's Department, Capital Works Unit (1992). Value Management, Sydney.

PMI (1987). *Project Management Body of Knowledge.* Project Management Institute, New York.

Quinlivan-Hall, D. & Renner, P. (1994). *In Search of Solutions.* Toppan/Pfeiffer, Singapore.

RTA (1992). *Project Management Guide.* Roads and Traffic Authority, Sydney.

United Nations Industrial Development Organisation (UNIDO), (1978). Manual for the Preparation of Industrial Feasibility Studies, New York.

WAPMA (1990). *Terms of Engagement for Project Management Services*. Western Australian Project Management Association, Perth.

Subject Index